CONFRONTING CLIMATE GRIDLOCK

DANIEL S. COHAN

Foreword by Michael E. Webber

Confronting Climate Gridlock

How Diplomacy, Technology,
and Policy Can Unlock a Clean
Energy Future

Yale UNIVERSITY PRESS

NEW HAVEN AND LONDON

Published with assistance from the foundation established in memory of
Philip Hamilton McMillan of the Class of 1894, Yale College.

Yale University Press books may be purchased in quantity
for educational, business, or promotional use. For
information, please e-mail sales.press@yale.edu (U.S. office)
or sales@yaleup.co.uk (U.K. office).

Set in Janson and Futura type by IDS Infotech, Ltd.
Printed in the United States of America.

ISBN 978-0-300-25167-8 (hardcover : alk. paper)
Library of Congress Control Number: 2021942326

A catalogue record for this book is available from the British Library.

This paper meets the requirements of ANSI/NISO Z39.48-
1992 (Permanence of Paper).

10 9 8 7 6 5 4 3 2 1

To Mira and Jacob,
For the climate you will inherit

. . . And if not now, when?
—Hillel

CONTENTS

Foreword by Michael E. Webber   ix

Preface   xiii

List of Abbreviations   xvii

one Why Climate Gridlock?   1

two The Road to Paris   12

three The Road from Paris   37

four Pillars of Decarbonization   57

five Decarbonizing Electricity   73

six Power Shift   103

seven Going Negative   126

eight Confronting Policy Gridlock   140

Notes   167

Acknowledgments   223

Index   225

FOREWORD

For more than a century, energy has enabled globalization. Modern forms of energy such as coal, diesel, or jet fuel have been responsible for moving most of the world's goods around by railways, shipping lanes, and flight paths. This globalization means we can have modern luxuries such as fresh fruits that are always in season no matter the time of year. We also have access to inexpensive manufactured goods from halfway around the planet that make our lives comfortable and enjoyable.

Energy is itself one of those goods that is moved around. The world's demand for coal, oil, wood pellets, natural gas, and refined fuels spans oceans and continents. The disparate locations of energy exporters and importers connect us all in ways that bridge culture, language, religion, and geography.

What we have come to realize very sharply in the last few decades is that it's not just energy itself as a commodity of and fuel for trans-oceanic supply chains that define its global nature: the waste products likewise do so. Unlike environmental concerns from prior eras that were local in nature, greenhouse gas emissions have a global effect. In the past, environmental impact was limited in geographic scope. Water contamination would happen from a nearby mine. Air pollution would cause asthma in the factory town or acid rain in a neighboring state. But because greenhouse gases like carbon dioxide are long-lived and stable and mix rather uniformly in the atmosphere, climate change happens on a worldwide basis.

This phenomenon that emissions from discrete smokestacks or tail-pipes inflict global impact brings with it an inequitable situation in which the benefits of energy use are highly concentrated to the pro-ducers and consumers, but the downsides of its pollution are now shared by all. Even worse, the downsides are not shared equally. The people who will suffer the most from climate change have not been born yet, and when they are, they likely will be in countries with much lower energy use, emissions, and economic activity per capita than the Americans, Europeans, Japanese, and Australians, who are all at the high end of the spectrum.

How do we increase access to energy for those 1 billion people who will suffer from climate change but do not have modern lifelines such as electricity, piped water, or sanitation? Can we increase their access while decreasing the global climate effect of the other 7+ billion who already have access? How do we change an industry active in every country and whose impacts, no longer isolated, are endured worldwide?

The answers will not be easy and, more important, will not be just technical in nature. They will include human dimensions such as policy, international relationships, markets, culture, and behaviors. It is in that multidimensional vein that Professor Daniel Cohan's book argues inaction on climate can be overcome.

With all roads leading to or from Paris on international climate agreements and various pathways available to zero (or negative!) emissions, we need to see through the fog and tune out the noise. Cohan helps us do so. There are many important factors to consider for a problem this nuanced that has been in the making since the second half of the nineteenth century. But among them a few key lessons stand out. Policy matters. Zero isn't enough. The grid will become more important than ever. The word "gridlock" even has part of the answer—grid—embedded inside. Sometimes the hardest solutions to see are the ones right in front of us.

It is important that climate change is no longer some abstract future bogeyman we might need to think about someday. It is here, and its effects already reveal themselves in haunting ways. In Texas, where Cohan and I both live, we endure wildfires, droughts, floods, hurricanes, tornados, heat waves, and arctic freezes. The growing frequency and intensity of these episodes are a stark reminder of the risks we face.

Though climate change is global in nature, Cohan shows us that the United States has a special role to play. As the largest cumulative emitter of greenhouse gases since the Industrial Revolution, the United States, with its economic might and unique geopolitical position, has a special opportunity and responsibility to take action. The United States isn't the only actor that matters, but forward progress is more likely with American participation. And if the United States can successfully wrestle with its own climate legacy, it will help reveal a pathway forward that can gain traction elsewhere.

Cohan maps out a way forward with an eye toward problem-solving rather than finger-pointing, blaming, or shaming. The inspiration and skills that gave us the benefits of modern energy—necessity as the mother of invention combined with innovation, geology, chemistry, physics, engineering, enabling policies, and modern market designs— might also be what is needed to get us out of this mess.

In the end, if the United States steps up, we can solve this global challenge. But it's going to take a while, so we better get started.

Michael E. Webber

PREFACE

This book was born in the eye of a hurricane—several, in fact. Hurricane Ike knocked out my power as a birthday greeting in 2008. Then, in 2015, Hurricane Patricia set records with its 345 kilometer per hour (215 mile per hour) winds over the eastern Pacific before its remnants drenched Houston. Two devastating storms flooded Houston the following two years before Hurricane Harvey shattered continental U.S. rain records in 2018, leaving me mucking out friends' homes while mine was fortunately spared.

I had returned to my home state of Texas in 2006 to study air pollution, not hurricanes. Houston's air pollution had drawn attention during Governor George W. Bush's presidential race, when he was derided for letting Houston dethrone Los Angeles as the smog capital of the country. By the time I arrived, Houston had returned that unwelcome title, and our air has been getting cleaner ever since. Most people scarcely notice the scrubbers, catalytic converters, and other technologies that protect our air quality.

The global warming that is fueling record hurricanes here and other disasters around the world defies straightforward solutions. Carbon dioxide accumulates in the atmosphere globally and lingers for centuries, even as other air pollution rains out regionally within days. Air pollution is the bigger killer today, responsible for several million deaths per year globally. But it is climate change that poses an existential threat to life as we know it in the decades ahead. Efforts to control it have for too long been gridlocked.

As an atmospheric scientist, I could lecture for hours about the chemistry and physics of air pollution and climate change, until my students slump in their seats or nod off on Zoom. But what excites me most as a professor of environmental engineering is the quest for solutions. No environmental challenge needs solutions more than global warming. Such solutions must inherently transcend the realms of diplomacy, technology, and policy that are the focus of this book.

Confronting climate gridlock will require tackling emissions around the world. Emissions anywhere warm the planet everywhere. But emissions mitigation is less a global challenge than an amalgam of national ones. The power systems, vehicles, buildings, industries, and agriculture that are the dominant sources of emissions are all regulated mostly on national or smaller scales. Thus, climate gridlock must ultimately be confronted on national scales. Diplomacy and technology transcend and transform those national challenges worldwide, but politics and policies are particular to each nation. This book explores diplomacy and technologies that can be applied worldwide, but it gives particular attention to the politics and policies of the United States.

The United States is by no means representative of nations confronting climate gridlock. America is exceptional, for better and worse. Our combination of diplomatic clout, technological prowess, natural resource abundance, and political dysfunction is uniquely our own. But the actions of the United States are distinctly pivotal to the progress that can be achieved in other nations. The United States ranks second only to China in emissions today and tops the world in emissions historically. With the world's largest economy, our markets influence manufacturing worldwide. No other country is better positioned to develop technologies crucial to making clean energy attractive and affordable, even if they are ultimately manufactured elsewhere and deployed around the world. And no other country has wavered so prominently in its support for climate treaties or seen the solutions and even the science of climate change become so politicized. Our ability to lead or obstruct progress by others will depend on whether we commit more fully to action in the decade ahead.

This is a book about solutions. I touch only briefly on the causes and impacts of climate change, taking as a given the scientific consensus that emissions of greenhouse gases are warming our planet at great detriment to society and nature. Instead, I focus on the three keys— diplomacy, technology, and policy—that will be needed to unlock climate gridlock in the United States and catalyze progress abroad.

After introducing the challenge of climate gridlock (Chapter 1), I discuss the history and future directions for climate diplomacy

(Chapters 2 and 3). Chapters 4–7 explore technological pursuits of clean energy, beginning with an overall framing and then focusing particularly on efficiency (Chapter 4), electricity (Chapter 5), electrification and clean fuels (Chapter 6), and negative emissions technologies and geoengineering (Chapter 7). Discussion of associated policies is interwoven into the technology chapters and then expanded in the context of history, political science, and emerging trends to explore potential pathways forward in Chapter 8. What will emerge is an intricate portrayal of daunting challenges and hopeful opportunities for confronting climate gridlock and unlocking a clean energy future for the United States and beyond.

ABBREVIATIONS

ACEEE	American Council for an Energy Efficient Economy
ARPA-E	Advanced Research Projects Agency—Energy
BECCS	Bioenergy with carbon capture and storage
CFCs	Chlorofluorocarbons
COP	Conference of the Parties (annual climate talks, including COP 21 in Paris in 2015)
DDPP	Deep Decarbonization Pathways Project
DOE	U.S. Department of Energy
EIA	Energy Information Administration
EPA	U.S. Environmental Protection Agency
FERC	Federal Energy Regulatory Commission
HFCs	Hydrofluorocarbons
IGCC	Integrated gasification and combined cycle
IPCC	Intergovernmental Panel on Climate Change
NDC	Nationally Determined Contribution under the Paris Agreement
NREL	National Renewable Energy Laboratory
R&D	Research and development
RD&D	Research, development, demonstration, and deployment
RMI	Rocky Mountain Institute
UNFCCC	United Nations Framework Convention on Climate Change
WTO	World Trade Organization

CONFRONTING CLIMATE GRIDLOCK

one WHY CLIMATE GRIDLOCK?

The quest to slow global warming is for now gridlocked. After uniting under the Paris Agreement of 2015 to set global targets for temperature and net-zero emissions of climate-warming gases, countries have failed to pursue the national actions and international collaborations needed to achieve them. Technologies for clean energy continue to improve but have barely dented the dominance of fossil fuels. American policies have failed to adequately mitigate emissions here or demonstrate leadership abroad. All of this has left emissions near record highs globally and declining only slowly domestically, even as drastic cuts are needed to stabilize the climate.[1]

Climate change is an inherently global problem. Carbon dioxide emitted anywhere warms the climate everywhere, wafting through the air for centuries alongside shorter-lived greenhouse gases. Warming is already leading to rising seas, stronger hurricanes and floods, more intense droughts and wildfires, proliferating pests, and unpredictable shifts in winds and ocean currents. No community or ecosystem on Earth is immune from these hazards, with the greatest risks falling on those least able to confront them.

The climate problem may be global, but solutions must arise mostly at the national level and be woven together with international cooperation. Despite the ubiquity of fossil fuels as globally traded commodities, their use isn't truly global. Instead, fossil fuel use is rooted in energy systems that are more or less national in scale. Power plants operate in regionally managed grids. Vehicles and industries are regulated mostly by national emission standards. Building codes are set by state and local governments and influenced by federal policy. Thus, actions at national or smaller scales matter most.

Some nations' actions matter more than others. Of the nearly two hundred parties to the Paris Agreement, just seven of them— China, the United States, the European Union, India, the Russian

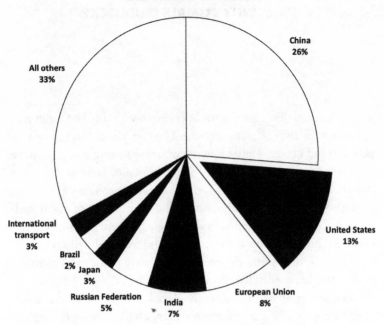

Figure 1. Greenhouse gas emissions by country in 2018 (Plotted by the author with data from Table B.1 of J.G.J. Olivier and J.A.H.W. Peters, "Trends in Global CO₂ and Total Greenhouse Gas Emissions: 2019 Report" [PBL Netherlands Environmental Assessment Agency, May 2020])

Federation, Japan, and Brazil, plus international transport that is largely between them—emit two-thirds of all greenhouse gases (Figure 1). Thus, confronting climate gridlock globally depends mostly on a handful of parties unlocking the barriers to their own progress on clean energy and establishing an international playing field that fosters decarbonization worldwide.[2]

The relative importance of the United States among these heavyweights is in some ways waning. As emissions from China and developing countries soared, the U.S. share of global emissions declined from 20 percent in 2005 to 13 percent in 2018. American clout in climate diplomacy has dwindled too, as other nations remained committed to the Paris Agreement even as the United States bobbed in and out. Other nations have faster-growing populations and economies, more ambitious policies, or leadership in certain key technologies.[3]

Nevertheless, the United States retains a uniquely pivotal role in the climate problem and its solutions. The country is the largest contributor to cumulative emissions historically, and its emissions remain nearly twice as large as any other country besides China. Its fickleness politically has made the United States the leading wild card in international diplomacy. Europe has set a better example with steadier and stronger policies but has often soldiered ahead without persuading others to follow. Once the United States does act, it usually insists on others acting too, making it more likely to draw followers.

U.S. universities, corporations, and national laboratories lead the world in developing the technologies needed for a clean energy transition. No other country has a broader array of corporations, banks, venture capital firms, foundations, and government agencies to fund the deployment of those technologies. Enhancing the performance and shrinking the costs of clean energy technologies here can make them more attractive and affordable abroad. "If we can develop technologies that can be used around the world . . . then American innovation can help decarbonize other economies," said Varun Sivaram, an expert on clean energy.[4]

I therefore argue in this book that breaking through climate gridlock in the United States is pivotal to confronting climate gridlock globally. That U.S. breakthrough will require grasping and aligning three keys—diplomacy that creates an international playing field that motivates U.S. action and leverages it to accelerate action abroad; technologies that make emission reductions achievable and affordable across all sectors of the energy economy; and policies that accelerate the development and adoption of those technologies domestically and abroad. Each of the keys remains ungrasped, but underappreciated trends are bringing them closer within reach than ever before.

None of the three keys can be fully grasped without the others. Domestic policy needs diplomacy to tilt the international playing field to favor decarbonization and technologies to make decarbonization practical and affordable. Diplomacy needs domestic policies for credibility and technologies to make clean energy desirable worldwide.

Technologies need funding for research and development and policies to promote their deployment. Only by grasping and aligning the three keys of diplomacy, technology, and policy together can we unlock a path to a clean energy future domestically and abroad.

A Measure of Degrees

Success or failure in confronting climate gridlock will ultimately be gauged by how high temperatures rise. The Paris Agreement committed the world to hold the increase in the global average temperature "well below 2°C" above "pre-industrial levels," and to "pursu[e] efforts" for 1.5°C. Thus, it is worth taking a moment to understand the implications of these and other temperature thresholds and the importance of staying within them.

The "pre-industrial" baseline against which warming is measured was left undefined by the Paris Agreement. Ice cores, corals, and tree rings all help scientists approximate temperatures deep into the pre-industrial past. However, high-quality thermometer measurements were not available over enough of the globe until the 1880s to reliably estimate a global average temperature. Setting a pre-industrial baseline around that time captures nearly all of the warming from fossil fuels, since emissions across the entire previous century were less than half what the world emits in a single year today.[5]

Temperatures have now soared 1.2°C (2.2°F) above their 1880–1900 levels, with most of that rise coming since the 1950s (Figure 2). The pace of warming is accelerating, reaching 0.3°C per decade in the 2010s. At that pace, we would enter the 1.5 to 2°C Paris target range during the 2030s and break through it by midcentury.[6]

Each degree of warming may not seem like much, even if we multiply it by 1.8 for those of us who think in Fahrenheit. (I'll use Celsius for the remainder of this book, for consistency with the language of treaties and science.) After all, temperatures swing by tens of degrees each time a cold front or a heat wave rolls through. But weather isn't climate. Seemingly small shifts in climate have profound consequences. Also, global averages understate the warming that most people will experience. Land warms faster than oceans, which can

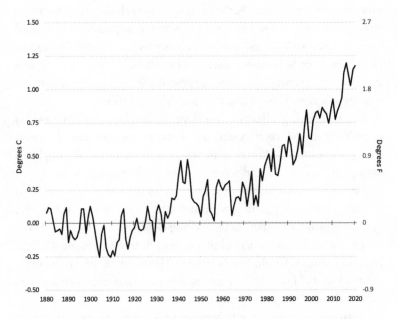

Figure 2. Global average temperature relative to 1880–1900 (Plotted by the author with data for annual global land and ocean temperatures from NOAA National Centers for Environmental Information, Climate at a Glance: Global Time Series, retrieved on March 16, 2021, from www.ncdc. noaa.gov/cag)

dissipate heat to greater depths. Nights warm faster than days, since greenhouse gases absorb Earth's radiation around the clock but particle pollution reflects sunlight only during daytime. Thus, over land, the hottest days are expected to warm twice as fast and the coldest nights three times as fast as the global averages.[7]

Most warming since the Industrial Revolution has been initiated by carbon dioxide, with smaller contributions from less abundant but more potent greenhouse gases such as methane, nitrous oxide, chlorofluorocarbons (CFCs), and hydrofluorocarbons (HFCs). Greenhouse gases allow visible sunlight to pass through but absorb some of the infrared heat radiated by the Earth's surface. The gases release some of that heat back toward the surface, warming the planet.

Human activities, especially the burning of fossil fuels, emit enormous amounts of these gases. The resulting warming is amplified by natural "positive feedback cycles" that cascade across the climate. For example, warmer air holds more water vapor, which is itself a greenhouse gas. Melting glaciers and sea ice expose darker land and sea surfaces that absorb more sunlight, further amplifying warming.

The amplifying effects of positive feedback cycles have swung Earth's climate between ice ages and warm periods for eons. Previous swings were kicked off mainly by wobbles in Earth's orbit, amplified by positive feedbacks and modulated by oscillations in solar output and volcanic activity. This time around, no natural kick can explain warming that is happening with unprecedented speed. In fact, the current position of Earth's orbit, passing closest to the Sun in January, should be holding us in a cool phase, since it minimizes the sunlight available to melt ice during Northern Hemisphere summers. Only human-emitted greenhouse gases, amplified by natural feedbacks, can explain the warming that is underway.

Warming is for now partially offset by a haze of particle pollution that reflects sunlight away and brightens clouds. Despite its cooling effect, particle pollution causes millions of deaths per year. Its cooling influence will fade as emissions are controlled, since particles rain out within weeks but long-lived greenhouse gases accumulate for decades.

In sum, the warming we have experienced so far is due almost entirely to manmade greenhouse gases, amplified by natural feedbacks and dampened temporarily by cooling from particles. Though the sensitivity of temperatures to emissions remains uncertain, as a rough approximation it takes around 200 billion metric tons (gigatons) of carbon dioxide, or nearly 500 billion barrels of oil, to warm the planet by one-tenth of a degree Celsius. As the world emits more than 40 gigatons of carbon dioxide per year and other greenhouse gases continue to rise, temperatures are rising accordingly.[8]

Each degree of warming yields countless impacts across the planet. Here in my hometown of Houston, the first 1°C of warming more than tripled our likelihood of experiencing deluges on the scale of Hurricane Harvey, which smashed rainfall records while inundating thousands of homes. Research has shown that the majority of the economic damage from that storm can be attributed to human-induced climate change. The rainiest days in New England and the Midwest are now about 40 percent rainier than they were a century ago. Droughts are becoming more severe, as manifest in the epic wildfires that have ravaged western states. Coral reefs are beginning to succumb to warmer and more acidic oceans, as those of us who scuba dive can already see. Arctic sea ice has shrunk to just over half of its former extent, while Greenland is losing 270 gigatons of ice per year, enough to cover the entire island in a 13-centimeter puddle of water.[9]

Even halting warming at 1.5°C would leave substantial impacts. Hurricanes, floods, and heat waves would all continue to intensify. Agriculture would be threatened by droughts in some regions and extreme storms in others. Sea ice and coral reefs would dwindle but not disappear. Sea levels would rise at least another 30 centimeters (1 foot) by 2100.[10]

Risks worsen between 1.5°C and 2°C of warming. That extra half-degree would exacerbate droughts and expose an additional 420 million people globally to extreme heat waves. It would also add an extra 10 centimeters of sea level rise by 2100, and an extra 30 centimeters of rise by 2300. Sea ice would disappear from the Arctic Ocean during some summers, and virtually all of the world's tropical coral reefs would be destroyed.[11]

Although human responses to climate change are difficult to predict, social scientists warn of risks ahead. A meta-analysis found that warming and extreme rainfall significantly increase rates of interpersonal violence and intergroup conflict. High temperatures tend to undermine economic productivity and worsen inequality. Economists estimate that an extra 1°C of warming would cost more than 1 percent of gross domestic product (GDP) in the United States, with Texas and southern states hit the hardest.[12]

Warming beyond 2°C heightens the risks of surpassing "tipping point" thresholds that could trigger irreversible changes in Earth systems. Those changes include a slowdown in ocean circulation, die off of forests, thawing of permafrost, and destabilization of ice sheets that could set off a cascade of follow-on impacts and instability. Scientists initially assumed that tipping points would not be reached until warming of 5°C or so, but more recent research has identified risks closer to 2°C.

"Once we stray into the 2 degrees or more warming, we get into the tens of percent probability for hitting some of these tipping points," said Tim Lenton, a climate scientist from the University of Exeter who studies tipping points. "Then, if we carry on business as usual, it's going to be more likely than not that we'll hit at least one tipping point." That could trigger what climate scientist Will Steffen calls a "hothouse Earth," featuring a "domino-like cascade" of further warming and disruptions. Such scenarios pose an uncertain risk and would take centuries to unfold, but they give added reason to hold warming well below 2°C.[13]

Three degrees of warming would push us beyond conditions that the Earth has experienced in millions of years, so the hazards are even more unpredictable. "The greater the change, the greater the risk," climate scientist Katharine Hayhoe of Texas Tech University told me. "We are truly conducting an unprecedented experiment with our planet." Agriculture would be devastated, ecosystems transformed, and some already hot regions would become virtually uninhabitable. Over vast regions including southern Europe and the southern and central United States, what are now considered 1-in-100-year droughts would begin occurring every two to five years. Hurricanes and typhoons would more often reach catastrophic intensity, with storm surges compounded by rising seas.[14]

Actions already being taken under the Paris Agreement and beyond make it unlikely that temperatures will climb much beyond 3°C this century. But these steps are not enough to prevent temperatures from continuing to rise in the following century. Warming of 4°C and beyond would yield almost unimaginable consequences for our descendants to bear.

Warming will cease only after emissions fall into a net-zero balance with the sinks that remove them. As in a bathtub with a tiny drain and a gushing faucet, slowing the flow of greenhouse gases into the atmosphere is not enough to stop their level from rising. It would merely slow the rise and the warming that comes with it. Since carbon dioxide remains in the atmosphere for centuries, the emissions we release today lock in warming for generations to come.

Exactly when we must reach a net-zero balance between sources and sinks is open to debate. Policy analysts and economists have developed a wide array of scenarios for emissions and sinks. Running those scenarios through different climate models yields different estimates of warming. Despite the uncertainties, most scenarios modeled to hold warming to 1.5°C would require cutting emissions in half by 2030 and bringing carbon dioxide into a net-zero balance between sources and sinks globally between 2045 and 2055, with net-negative emissions thereafter. For 2°C, scenarios vary more broadly, with most requiring roughly 25 percent reductions by 2030 and net-zero carbon dioxide globally sometime between 2060 and 2080. Unfortunately, projections from the United Nations Environment Programme suggest that even if countries uphold their Paris commitments, emissions will stay flat through 2030. That leaves an "emissions gap" of roughly 15 billion tons per year from a 2°C pace, and twice as much from a 1.5°C pace, that must be closed by 2030.[15]

A strong case can be made that the United States should cut emissions faster than other countries to help the world achieve these global targets. The United States has emitted far more carbon dioxide cumulatively than any other nation in history, giving it an outsized responsibility for warming to date. Our relative wealth enhances our capacity to invest in clean energy and negative emissions technologies. Abundant land and natural resources offer exceptional opportunities to harness renewable energy, harvest biomass, plant trees, and store carbon underground. Thus, net-zero carbon dioxide by 2050 is about as lax a target as we can give ourselves to be consistent with the Paris Agreement's global goal of aiming for 1.5°C and holding warming well below 2°C. That will require cutting emissions with unprecedented speed, halving them each decade while scaling up sinks.

Difficult as net-zero carbon dioxide emissions may sound, it is doable. In fact, separate teams of experts from the National Academies, Zero Carbon Consortium, Princeton University, Center for Climate and Energy Solutions, and Energy Innovation have all concluded that the United States could achieve net zero by 2050 at a cost of just a fraction of a percent of GDP while yielding substantial co-benefits for health and other factors. Stunning advances in technologies have made that pursuit affordable, but only with breakthroughs in diplomacy and policy can it be achieved.[16]

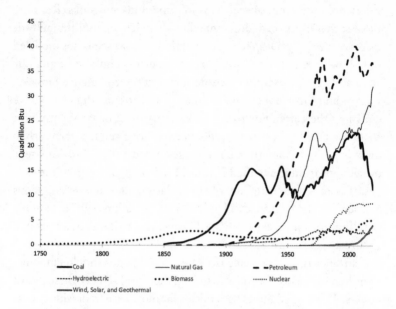

Figure 3. U.S. primary energy consumption by fuel. Primary energy consumption overstates the importance of fossil fuels like coal relative to renewable sources of electricity, since only about one-third of the energy content of coal is converted to electricity. (Plotted by the author with data from U.S. Energy Information Administration, "Monthly Energy Review," Tables 1.2 and D1, accessed July 23, 2020)

Unfortunately, slowness has been the hallmark of energy transitions historically. Pre-industrial civilizations were for millennia powered mostly by renewable resources: people, animals, water, and wood (Figure 3). It took over a century after the Industrial Revolution for coal to surpass wood as the leading source of energy. Nearly another century passed before oil and gas overtook coal. In other words, it took two hundred years for coal, oil, and gas to carbonize the energy economies of the world. "It took enormous amounts of effort to carbonize the energy system, so there's no way that it's not going to take enormous amounts of effort to decarbonize the energy system," said Matto Mildenberger, a scholar of climate policy at the University of

California, Santa Barbara. Aiming for net zero soon enough to meet the Paris targets will require us to achieve that decarbonization in just a few decades. As former president of the Climate Policy Center Rafe Pomerance told me, "This is the largest task that the world has ever had to undertake—to replace our carbon-based energy systems in sufficient time."[17]

Given the global nature of the climate challenge, I'll start with how diplomacy can catalyze global efforts before turning to the technologies and policies needed to unlock a clean energy future in the United States and beyond.

two THE ROAD TO PARIS

Banging his green gavel, French foreign minister Laurent Fabius was nearly drowned out by applause as he announced in French, "The Paris accord is adopted." Delegates from around the world rose to their feet as summit leaders clasped hands in jubilation on stage. After two weeks of negotiations and years of preparation, much of it coaxed along by Fabius himself, a consensus had been reached. United Nations Secretary General Ban Ki-moon called it a "monumental triumph." President Barack Obama hailed the agreement as "the best chance we have to save the one planet we've got."[1]

The 2015 accord established twin goals—a limit on warming and a pursuit of net-zero emissions—that have shaped climate diplomacy ever since. The accord also marked an armistice in a decades-long battle of ideas over how the international community should address climate change. At essence, that battle revolved around two questions: Should emissions limits be set "top-down" by internationally negotiated mandates or "bottom-up" by voluntary national pledges? And what are the "differentiated responsibilities" of wealthy "developed" nations and poorer "developing" ones? As shorthand, I'll call these the "how" and "who" questions of climate diplomacy. Although President Donald Trump denounced the Paris Agreement as "very unfair, at the highest level, to the United States," it was actually American ideas pursued across a quarter century that largely prevailed on each question. Looking back on the road that brought us to Paris provides crucial insights into what's needed to seize the international key to confronting climate gridlock.[2]

Lessons from Ozone Diplomacy

As environmental activism began to blossom after the first Earth Day in 1970, climate change barely ranked among the world's top environmental priorities. When the United Nations held its first international environmental conference in Stockholm in 1972, the resulting conference report waited until the seventieth of its 109 recommendations to blandly recommend "that Governments be mindful of activities in which there is an appreciable risk of effects on climate." It was not until 1979 that the first World Climate Conference was held in Geneva and the National Research Council issued its first scientific assessment of carbon dioxide and climate. Instead, it was ozone depletion that first captured the world's attention as a global atmospheric concern.

In a paper published in 1974 that later helped earn them the Nobel Prize in Chemistry, scientists Mario Molina and F. Sherwood Rowland discovered that chlorofluorocarbons (CFCs) can be split apart by ultraviolet sunlight in the stratosphere, releasing chlorine that depletes the ozone layer. It took just two years for the National Academy of Sciences to confirm the threat and for Congress to authorize the Environmental Protection Agency (EPA) to regulate CFCs. By 1978, the EPA had banned the use of CFCs in aerosol spray cans, though other uses remained (Figure 4).[3]

Ozone depletion morphed from a theoretical threat into an observed reality in 1984, when a team of British Antarctic Survey scientists announced its discovery of what would soon become known as the "ozone hole" over Antarctica. The British team had been observing steep declines in ozone above its Antarctic station each October since the late 1970s, but the scientists had dismissed these findings as a malfunction of their aging instrument. After all, how could ozone be declining so fast if NASA satellite data were not showing such a trend? Little did they know that NASA had been filtering out low measurements, with a computer algorithm designed to reject data that seemed impossibly low. In other words, both the British team on the ground and American satellites in space were blinding themselves to ozone depletion in between their instruments.[4]

Global leaders responded to the emerging science with remarkable speed. Even before the British team and NASA scientists had formally published their findings, negotiators met in Vienna in 1985 to broker a treaty that set a framework for international action on ozone. Reconvening in Montreal in 1987, they crafted a protocol that set a timetable

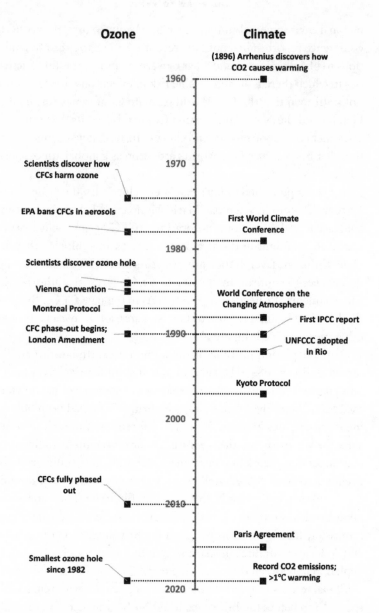

Figure 4. Timeline of science and diplomacy for stratospheric ozone and climate change

for a mandatory phase-out of CFCs and other ozone-destroying chemicals. Now ratified by 197 countries, the Montreal Protocol, along with subsequent reinforcing amendments, has eliminated 99 percent of emissions and has been called the "most successful international environmental treaty in history." The EPA expects that the protocol will eventually avert 45 million cataracts and 1.6 million deaths from skin cancer in the United States alone. As stratospheric ozone continues to rebound, scientists expect it to recover to its natural levels by the middle of the century. Though aimed at ozone, the Montreal Protocol has averted more warming than any other treaty, since CFCs are incredibly potent at warming the atmosphere.[5]

Unlike the partisan politics that would later plague climate policy, the Montreal Protocol was signed by President Ronald Reagan and ratified unanimously by the U.S. Senate in 1988. Its provisions were incorporated into the Clean Air Act through amendments signed by another Republican president, George H. W. Bush. Although the chemical giant Du Pont initially spearheaded an industry alliance that lobbied against action, by the late 1980s it had swung around to endorsing the Montreal Protocol. Chemical companies foresaw greater profits from producing more recently patented replacement compounds than generic CFCs. Manufacturers soon substituted more ozone-friendly compounds into their spray cans, refrigerators, and foams.[6]

The success of the Montreal Protocol for ozone would seem to make it a template for climate diplomacy. But ozone depletion and climate change are less similar than their global scopes would suggest. CFCs and carbon dioxide both accumulate globally in the atmosphere, but the gases arise from very different sources. CFCs were produced mainly as niche products by a handful of chemical giants operating in a limited number of countries. Du Pont alone controlled 27 percent of the global market, yet it derived just 2 percent of its sales from CFCs. Consumers barely noticed as CFCs were replaced by substitutes that barely affected the performance or cost of refrigerators and other products. All of this made CFCs a ready target for a global phase-out imposed on a limited number of manufacturers.[7]

By contrast, carbon dioxide is an inevitable byproduct of a fossil-fueled economy. Fossil fuels still supply roughly 80 percent of the

United States' and the world's energy needs, percentages that have barely budged over the past three decades. Nearly every carbon atom from a combusted fossil fuel combines with two heavier oxygen atoms from the air. Thus, each ton of fossil fuel that we burn adds more than two tons of climate-warming carbon dioxide to the atmosphere. That averages out to nearly five tons of carbon dioxide, roughly the weight of two pickup trucks, for every person on Earth each year. Americans emit several times more per person. Transitioning away from fossil fuels requires redesigning the vehicles, power systems, factories, and buildings that rely on them. Hundreds of the world's largest corporations and countless smaller businesses owe their existence to producing, processing, or transporting fossil fuels, and just about every person on Earth consumes them. "Montreal was about pollutants that were microscopically important to the economy compared to carbon dioxide," former U.S. climate negotiator Susan Biniaz told me. Trying to apply a Montreal Protocol–style approach of globally set emissions limits to greenhouse gases soon led climate diplomacy into gridlock.[8]

The Road to Rio

Soon after the signing of the Montreal Protocol for ozone, climate change leaped to the forefront of environmental concerns both globally and domestically in 1988. President Ronald Reagan and the Soviet Union's Mikhail Gorbachev signed a joint statement that June pledging to cooperate on addressing climate change. Later that month, the World Conference on the Changing Atmosphere, convened by the prime ministers of Canada and Norway in Toronto, recommended a 20 percent cut in global carbon dioxide emissions. Bipartisan bills in the House and Senate called for the United States to seek an international climate agreement. An El Niño event stewing in the Pacific Ocean pushed global temperatures to their hottest levels in history. As a record heat wave struck Washington that July, NASA scientist James Hansen testified to the Senate that manmade global warming was already underway. (That testimony and ensuing media coverage prompted me to choose global warming as the topic for my year-long middle school research project, sparking a fascination with the atmosphere that has driven my career to this day.) Later in 1988, Vice President George H. W. Bush campaigned on a vow to be an "environmental president," telling a Michigan rally: "Those who think we are powerless to do anything about the greenhouse effect forget about the 'White House effect'; as President, I intend to do something about it." By December, the United Nations and World Meteorological Organization had established the Intergovernmental Panel on Climate Change (IPCC), which to this day remains the world's leading authority on climate science.[9]

Two years later, President Bush pushed through two signature environmental achievements. Domestically, Congress passed amendments to the Clean Air Act that strengthened air quality protections and created a cap-and-trade system to tackle acid rain. Internationally, diplomats brokered the London Amendments to the Montreal Protocol that accelerated the phaseout of CFCs. President Bush also hosted a global conference on climate change, a prelude to climate talks at the Earth Summit two years later.[10]

However, as a recession dragged into 1991, economic recovery took priority over environmental protection. Also, the fall of the Berlin Wall and defeat of Saddam Hussein lessened the administration's motivation to woo potential allies through climate policy. Thus, by the time negotiations began ahead of the 1992 Earth Summit in Rio de Janeiro, President Bush had hardened his opposition to mandating emissions cuts or extending foreign aid through a climate treaty. Several scholars have pointed to 1992 as the turning point when the United States shifted from being a leader to a laggard in environmental diplomacy.[11]

Meanwhile, most other countries were eager to act. Developing nations sought funding to help them pursue alternatives to fossil fuels, akin to the support they received for CFC substitutes under the Montreal Protocol. Wealthier nations coalesced around a target of limiting their greenhouse gas emissions to 1990 levels by 2000. That modest "top-down" target reflected a realization that fossil fuels would be more difficult to replace than CFCs, but it was clearly inadequate to halt warming. The IPCC's first scientific assessment, issued in 1990, had shown that far steeper cuts would be needed to curb climate change and that the Earth could warm 3°C by 2100 if emissions continued to rise unabated. Still, there was hope that initially modest targets could be strengthened as technologies improved, just as amendments to the Montreal Protocol had done for CFCs.[12]

Locked in a tough re-election fight and with his economic team warning that Rio presented a "bet your economy decision" after the 1990–1991 recession, President Bush threatened to boycott the Rio summit unless the treaty nixed mandates for emissions limits and foreign aid. With consensus required under United Nations protocol and the United States still by far the world's largest economy and emitter, Europe and developing countries conceded to the United States and abandoned their push for immediate mandates. However, they prevailed in writing key principles into the Rio treaty that set the stage for tougher action later. Most notably, the treaty committed the world to "stabilize greenhouse gas concentration in the atmosphere at a level that would prevent dangerous anthropogenic interference with the climate system." The treaty did not specify what that level was, nor did it define "dangerous" interference. Neither did it set an explicit goal of "net zero," although stabilizing greenhouse gases will ultimately require a net-zero balance between sources and sinks. Nevertheless, the treaty cast aside the "be mindful" banality of the Stockholm conference of 1972 and planted the seeds for the warming limit and net-zero target that would ultimately be adopted in Paris. President Bush eventually joined 116 other heads of state and delegates from 178 nations to finalize the treaty in June 1992. The Senate ratified it just four months later.[13]

The Rio treaty, formally known as the United Nations Framework Convention on Climate Change (UNFCCC), has been dubbed the "constitution for international action on climate change." It has served as the basis for annual "Conference of the Parties" (COP) talks ever since, punctuated by summits in Kyoto (COP 3) and Paris (COP 21). The treaty was approved unanimously and has now been adopted by every member state of the United Nations. Although the United States never ratified the Kyoto Protocol and President Trump temporarily withdrew the country from the Paris Agreement, the United

States has remained steadfast in its involvement in the overarching system that emerged from Rio.[14]

The Rio treaty provided only vague guidance as to "who" should act and "how." It stated that parties should act "on the basis of equity and in accordance with their common but differentiated responsibilities and respective capabilities," language that would later be mirrored in the Paris Agreement. But what is equitable? Defining "equity" and "differentiated responsibilities" is especially difficult for climate action, since countries may seek competitive advantage by doing as little as possible while pushing trade rivals to act. Rather than setting responsibilities for each country individually, the Rio treaty followed the Montreal Protocol's approach of grouping countries into broad categories. For Rio, that meant differentiating developed countries, comprising members of the Organisation for Economic Co-operation and Development and the former Soviet bloc, from everyone else. The developed countries were required to take the lead and craft plans to roll back their emissions to 1990 levels, though President Bush refused to accept a timetable for doing so. The treaty set no specific targets for emissions from developing countries, an omission that would plague climate diplomacy for the subsequent two decades.[15]

Despite its lack of mandates, the framework for climate diplomacy that emerged from Rio in 1992 was actually stronger than the one for ozone that had emerged from the Vienna Convention of 1985, which merely called for further research and talks. The Rio treaty explicitly called for emissions cuts, even if it didn't say how much or by whom. But after Vienna, it took just two years to broker the Montreal Protocol and another three years to negotiate the London Amendments, which required developing countries to follow the lead of developed countries in phasing out CFCs. Scientists have confirmed that CFC production has been nearly eliminated worldwide, atmospheric levels of CFCs are declining, and stratospheric ozone is beginning to recover. By 2019, the Antarctic ozone hole had shrunk to its smallest size since the early 1980s.[16]

By contrast, greenhouse gas emissions and warming accelerated after Rio. Between the Rio summit in 1992 and the Paris Summit in 2015, emissions soared 51 percent globally and far faster in developing

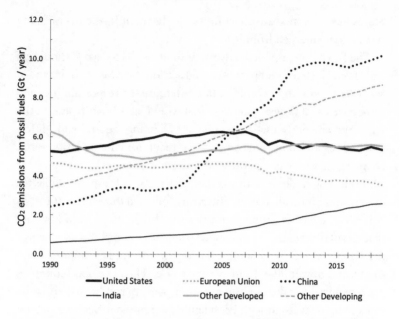

Figure 5. Carbon dioxide emissions from fossil fuels (gigatons/year) by region, 1990–2019 (Plotted by the author with data compiled by Lori Siegel of Climate Interactive from Friedlingstein et al., "Global Carbon Budget 2020," *Earth System Science Data* 12 [2020]: 3269–3340)

countries (Figure 5). Temperatures rose by 0.7°C during that time span. That's not entirely the fault of climate diplomacy. Fossil fuels form the foundation of regional and national energy systems, making them far more difficult to control than globally manufactured niche products like CFCs. Still, however difficult the challenge, the Rio treaty and its initial successors failed to solve it. In fact, climate diplomacy after Rio was largely a saga of missed opportunities, before a new paradigm emerged in Paris.[17]

From Rio to Kyoto

After the Rio treaty sidestepped disputes about the "how" and "who" of reining in emissions, those disputes took center stage in the two rounds of follow-up talks held in Berlin in 1995 and Geneva in 1996, which set the stage for the Kyoto Protocol. Whereas President Bush had forced negotiators to weaken the Rio treaty, the United States by this time was led by President Bill Clinton and Vice President Al Gore, who had made climate change his signature issue while representing Tennessee in Congress. However, their options were constrained after Republicans reclaimed the House and Senate in 1994. With a two-thirds vote in the Senate required to ratify any treaty, as few as thirty-four senators have virtual veto power over U.N.-sponsored climate negotiations, which operate based on international consensus.[18]

Aiming for a treaty that could win Senate approval as the Rio treaty had achieved, Clinton's climate negotiators initially adhered to three Bush-era positions: (1) emissions caps should be voluntary, with market-based mechanisms to achieve them flexibly; (2) developing countries should not be exempt from emissions constraints; and (3) foreign aid from the United States should be minimized. The first position created tensions with environmentalists and Europeans, who saw mandates as necessary to ensure progress. They pointed out that carbon dioxide emissions had continued to climb under the mandate-free Rio treaty, while CFC emissions were plummeting under the mandates of the Montreal Protocol. The latter two positions drew opposition from developing countries, which argued that their smaller historical responsibility and poorer resources should excuse them from immediate action and entitle them to financial support for subsequent action. The Montreal Protocol had done exactly that for CFC controls.[19]

Outnumbered at the Berlin summit in 1995, Gore relented to demands from Europe and developing countries that subsequent talks would focus on limiting emissions exclusively from developed countries. The deal, brokered by Germany's then–environment minister and later chancellor, Angela Merkel, gave negotiators two years to craft a treaty with specific emissions targets and timetables for developed countries alone. Developing countries, even rapidly industrializing ones such as China, would face no limits. Gore's acquiescence in letting some trade rivals off the hook may have clinched a deal in Berlin, but it infuriated Republicans in the Senate. The Senate Energy and Natural Resources Committee, chaired by Senator Frank Murkowski (R-Alaska), fired off a press release saying that American negotiators "got hoodwinked" and "should have walked away."[20]

Facing this blowback, the Clinton administration tried to renege on its concession. But developing countries refused to be included in the "who" of initial climate action. Meanwhile, Europeans and environmental groups, fearful that voluntary pledges would be inadequate, insisted that the "how" should be binding limits. At climate talks in Geneva in 1996, American negotiators conceded to those demands as well. Senator Murkowski again colorfully denounced Clinton's team for being outmaneuvered by trade rivals, saying, "If the climate change negotiations were strip poker, our negotiators would be down to their skivvies."[21]

Senate opposition to the U.S. concessions intensified in 1997 as the Kyoto talks drew closer. Senate minority leader Robert Byrd, representing coal-rich West Virginia, said negotiators made a "fundamental blunder" when they agreed to the Berlin Mandate, denouncing it as a "free pass" for developing nations. Together with Senator Chuck Hagel (R-Nebraska), he introduced a resolution opposing any treaty that did either of two things: set mandatory emissions limits for the United States without also setting them for developing countries, or risked harm to the U.S. economy. Recognizing that it would pass despite their opposition, climate hawks led by Senator John Kerry (D-Massachusetts) swung around to join in unanimous passage of the Byrd-Hagel resolution. That let them put a positive spin on the outcome, with Senator Patty Murray (D-Washington) claiming that the resolution "strengthens our bargaining position to ensure real, attainable standards that are established for developing countries, too."[22]

Such wishful thinking may have helped senators like Kerry and Murray save face. But the Byrd-Hagel resolution's twin conditions put the Clinton administration in a bind. A treaty that left out developing countries would be dead on arrival in the Senate. But most developing countries clung to the Berlin Mandate and refused to make even voluntary emissions commitments until after industrialized countries acted. Many emitted less than a tenth as much per person as the United States. Why should they put their fledgling economies at risk when fossil fuels could help lift their populations out of poverty? China was particularly adamant in refusing to limit its emissions. "China was very much an immovable object," recalled Melinda Kimble, who as a State Department official joined American and Japanese climate negotiators on fruitless visits to Beijing. "They wanted to bring as many of their billion people as possible out of poverty, and they thought their way of achieving this was the industrial revolution approach followed by the United States in the 19th century."[23]

By the time climate talks began in Kyoto in 1997, developing countries had cemented their opposition to immediate cuts, leaving wealthier countries to negotiate their own cuts in backroom deliberations. In the waning hours of the summit, Gore brokered a deal that split the difference between the 15 percent cuts sought by Europeans and the return to 1990 baseline levels sought by Americans. Under the deal, the United States and the European Union committed to cutting emissions by 7 and 8 percent, respectively, below 1990 levels by 2008–2012. Other targets were set for other developed countries, but none for developing ones. Though caps were said to be legally binding, no penalties were set for non-compliance.[24]

Knowing that caps for the United States but not developing countries would be a tough sell in the Senate, American negotiators won agreement for measures aimed at lowering costs and prodding developing countries to act. To enhance flexibility, wealthy countries were allowed to trade emissions allowances among themselves or buy them from emissions-cutting projects in developing countries. The latter approach was dubbed the Clean Development Mechanism. However, since developing country emissions were uncapped, it was often unclear whether projects funded under the Clean Development Mechanism yielded real cuts beyond the status quo.[25]

With the Kyoto Protocol signed but not ratified, the Clinton administration mobilized to address senators' twin concerns: the lack of mandates for developing countries and the cost of mandates for the United States. Diplomats tried to cajole individual nations such as Argentina and Kazakhstan to announce plans for action. But they faced a catch-22. "Our dilemma was that until we ratified Kyoto, getting developing countries on board . . . to take more action was a very heavy lift," Kimble told me. The Asian financial crisis made countries there especially reluctant to act. As for costs, Clinton administration officials tried to make the case that the Kyoto Protocol targets could be achieved affordably. But their claims were refuted by misinformation campaigns funded by industry-backed groups such as the Competitive Enterprise Institute and right-wing think tanks such as the Heritage Foundation, which warned of massive price spikes for electricity and gasoline. Meanwhile, campaigns to rally grassroots support for climate action were drowned out as the Monica Lewinsky scandal and then impeachment hearings dominated the headlines.[26]

Despite concerted efforts by the Clinton administration to persuade them, Republican senators refused to budge, letting the treaty languish without a vote. "The United States let the world down by agreeing to something that we should have known we could not deliver," said Nigel Purvis, who joined the State Department's climate team just after the Kyoto talks. He called the twin concessions made in Berlin and Geneva, acceding to no caps for developing countries and binding caps for developed ones, "fatal flaws" that ultimately doomed his team's efforts to win Senate backing for the treaty. Even without those flaws, it's unclear whether Republicans would have approved the Kyoto deal. "When the opposition is hiding behind one excuse after another, you can design the best system possible and they won't buy it," former deputy assistant secretary of state Rafe Pomerance told me.[27]

Campaigning in 2000, George W. Bush denounced the Kyoto Protocol as unfair to American industry, even as he called for caps on power plant emissions. After defeating the vice president who had brokered the Kyoto Protocol, President Bush showed initial glimmers of his father's environmental bent, appointing backers of climate action such as Secretary of State Colin Powell, EPA administrator Christine Todd Whitman, and Secretary of the Treasury Paul O'Neill. Powell directed officials at the State Department to seek out ways to renegotiate the Kyoto Protocol to make it palatable to more senators. Whitman traveled abroad to assure allies that the United States remained committed to reducing emissions. Back in Washington, O'Neill wrote memos to fellow Cabinet members and to President

Bush urging action. The leading senators from each party on the Senate Environment and Public Works Committee crafted a "grand bargain" to regulate carbon dioxide alongside other pollutants from power plants.[28]

Those administration officials and senators favoring climate action were soon outmaneuvered. Fossil fuel industry lobbyists and right-wing think tanks expressed alarm that President Bush might revive climate negotiations or follow through on his campaign pledge to regulate power plants. The lobbyists urged Senator Hagel and three fellow Republicans to issue a formal query to President Bush asking him to clarify his climate policy. The White House replied with a letter, purportedly from President Bush, that denounced the Kyoto Protocol as "an unfair and ineffective means of addressing global climate change . . . because it exempts 80 percent of the world . . . from compliance." The letter also renounced President Bush's campaign pledge to regulate power plant emissions, scuttling hopes for the Senate committee's "grand bargain" just days after Whitman had reiterated Bush's pledge. Later accounts by O'Neill and Whitman suggest that it was actually Vice President Dick Cheney, formerly the CEO of oil services giant Halliburton, who orchestrated both Hagel's letter and the reply that was issued under President Bush's name.[29]

Whoever wrote it, the White House letter extinguished hopes for U.S. ratification of the Kyoto Protocol. National security advisor Condoleezza Rice soon stated the obvious: "Kyoto is dead." The United States would remain on the sidelines as 192 other parties ratified the Kyoto Protocol and set up international markets to trade emissions.[30]

Evaluating Kyoto's impact is tricky. The countries that accepted limits under the protocol (that is, the former Soviet bloc and developed countries, minus the United States) cut their net emissions by 23 percent from 1990 to 2011. That's far better than the 5 percent aggregate reduction that the treaty sought. But progress was uneven. Nearly all of the overperformance came from the former Soviet-bloc countries. Their cuts have been dubbed "hot air," because economic collapse had already slashed their emissions below 1990 baselines before the Kyoto Protocol took effect. Europe roughly matched its commitments overall, with performance varying widely by country. Japan badly missed its target. Canada ultimately withdrew from the protocol after missing its target.[31]

The trouble is what Kyoto left untouched. Unbound by the protocol, U.S. emissions jumped 12 percent above their 1990 baseline by 2010, versus the 7 percent cut that Kyoto sought. China more than tripled its emissions, catapulting past the United States to become the world's leading polluter. Some other fast-growing countries such as South Korea more than doubled their emissions. Overall, global fossil fuel and industrial emissions of carbon dioxide jumped 46 percent from 1997 to 2012, more than one and a half times as fast as the growth rate of the prior fifteen years. On the other hand, compliance with the Kyoto Protocol led to innovations such as the European Union Emissions Trading System, which has overcome its initial struggles and is now a cornerstone of the bloc's push toward net zero.[32]

Copenhagen Accord

After Cheney's maneuvering pushed the United States to the sidelines, climate diplomacy entered what legal scholar Daniel Bodansky calls a "deep freeze." Europe set up markets to pursue its emissions targets, with little hope of bringing the world's two largest emitters, the United States and China, onboard. Annual climate talks came and went with little headway. The freeze began to thaw in Bali in 2007, when two paths for post-Kyoto climate action emerged. The first path sought to amend the Kyoto Protocol and set new emissions limits for the years ahead, continuing its approach of top-down mandates and aiming to bring aboard developing countries to the extent possible. The United States largely sat out those efforts as a nonparticipant in the Kyoto Protocol. The second path sought a new paradigm for climate policy. Delegates in Bali gave themselves until the December 2009 summit in Copenhagen to pursue each path and settle on a consensus route forward.[33]

Hopes for the 2009 talks soared with the election of Barack Obama. In his first address to Congress, President Obama called to "save our planet from the ravages of climate change." By July, President Obama had joined his fellow Group of 8 leaders in Italy in pledging to seek a deal for a 50 percent reduction in global greenhouse gas emissions by 2050, including an 80 percent reduction from developed countries. In other words, the "who" would be everyone, but wealthy countries would cut emissions more steeply.[34]

Fresh from receiving the Nobel Peace Prize in Oslo, President Obama arrived in Copenhagen touting "bold action at home" and a renewed commitment to leadership in climate diplomacy. With the economy mired in a deep recession, an economic stimulus bill had devoted $80 billion to clean energy. A House-passed cap-and-trade bill to cut emissions 83 percent by 2050 had not yet failed in the Senate. The EPA, with its newfound authority to address climate change under the Supreme Court's landmark 5–4 ruling in *Massachusetts v. EPA*, was preparing to regulate greenhouse gas emissions directly in case legislation failed.[35]

However, as protesters filled the surrounding streets demanding stronger action, talks inside the Copenhagen summit quickly broke down. In chronicling the talks, science journalist William Sweet wrote that the Danes "seemed way over their heads" managing the summit. The tiny country's leaders lacked the deft diplomacy that Fabius and his legendary corps of French diplomats would later use to forge compromise in Paris.[36]

But the deeper problem was the chasm between the positions of the negotiating parties. Poor and vulnerable countries sought binding emissions caps aimed at holding warming below 2°C and funding from wealthy nations to mitigate and adapt to climate change. U.S. negotiators knew that a treaty that forced the country to meet binding emissions limits or contribute massive amounts of foreign aid stood little chance of winning a two-thirds vote even in the Democrat-led Senate. They advocated a system of bottom-up national pledges, subject to rigorous reporting, that ultimately became the model for the Paris Agreement. China, which had recently surpassed the United States as the world's leading carbon emitter, refused to set a deadline for its emissions to peak, let alone decline. It also rejected rigorous rules for reporting. Europeans were eager to garner any deal they could get but were unable to bridge the chasms. Moreover, the Great

Recession had pushed climate change to the back burner as countries focused on reviving their economies.[37]

Against those headwinds, negotiators could cobble together only a three-page agreement in the waning hours of the summit. Known as the Copenhagen Accord, it was merely "acknowledged" rather than formally adopted by the conference. Even so, the accord marked a turning point. Of the two paths that had emerged from Bali, only the second, more open-ended path proceeded. The Kyoto path dead-ended in the accord's appendix, which had been intended to show emission limits for a second phase of the protocol. Instead, it held a blank table on an empty page. Countries instead vowed to issue their own national commitments, setting the stage for the Paris Agreement. A system of "bottom-up" nationally set pledges, advocated by Americans since Rio, had prevailed over "top-down" internationally set mandates. In return, developing countries won a pledge for $100 billion per year to help them mitigate and adapt to climate change. Despite the sticker price, the pledge was mostly hollow, since it included public or private or even "alternative sources of finance" and set no mechanism to ensure that the funding materialized. Some estimates suggest that funding reached only about half of the Copenhagen target in the ensuing decade.[38]

With its "accord" not even formally adopted, the Copenhagen talks left the international key to a climate breakthrough ungrasped. As Sweet put it, Copenhagen "epitomized all that had been wrong in two decades of climate diplomacy." A landmark treaty would ultimately have to wait until 2015 in Paris.[39]

Paris Agreement

The seeds of the Paris Agreement were sown long before delegates arrived in the French capital. The Bali talks in 2007 had charted two paths for climate diplomacy, and only the path of bottom-up pledges had continued onward from Copenhagen. At the 2011 climate talks in Durban, South Africa, countries agreed to pursue a legally binding deal by 2015 that would take effect in 2020. At the intervening summits, negotiators made their most crucial decision: each party would submit a "nationally determined contribution" setting forth its intended actions to mitigate climate change. The "how" of pledge-and-review had won out. The other American priority, that "who" would be everyone, prevailed as well. No longer would developing countries get a free pass. "The genius of Paris is that we've eliminated the idea that it's a two-tiered process," Kimble told me. "Under the Paris Agreement, everybody acts."[40]

Months before Fabius welcomed delegates to Paris, nearly every country had submitted its intended contribution. The most pivotal pre-summit breakthrough came when President Obama and President Xi Jinping of China reached an agreement in Beijing in November 2014 that outlined their intended national contributions, established cooperation on clean energy research and deployment, and vowed to seek common ground in Paris. Left to be decided in Paris were technical details as well as the wording of the agreement's twin goals.[41]

Observers of the Paris talks recall an unprecedented spirit of optimism and cooperation, following the tone set by French leaders. Terrorist attacks on the city two weeks before lent a sense of solidarity to the delegates as they met under tight security. Various countries, interest groups, and activists staked out conflicting positions in Paris, but a spirit of compromise prevailed. The Climate Action Network, representing nine hundred nongovernmental organizations, focused on advocacy rather than street protests like those that had roiled Copenhagen. Europe and developing countries pushed for national commitments to be made legally binding but backed down to American opposition, averting a Kyoto-style schism. Final proceedings were delayed ninety minutes as Americans forced through wording of "should" rather than "shall" to codify the non-binding nature of key passages. Regarding transparency, a broad coalition overcame Chinese resistance to rigorous reporting of emissions, though details were left to be resolved at subsequent COP talks. In terms of targets, most developing countries pushed for a rigorous $1.5°C$ warming limit and net-zero emissions by 2060–2080; most wealthier countries preferred $2°C$ and somewhat weaker long-term emissions targets. Negotiators met behind closed doors to paper over their differences.[42]

All parties could leave Paris pointing to a win. Americans may have been the biggest winners, prevailing on the "how" of non-binding, bottom-up pledges with rigorous reporting and the "who" of including all nations. As President Obama's top climate envoy Todd Stern recalls, many of the key provisions of the agreement "were literally our ideas." Europeans cheered the return of the Americans from their Kyoto isolation. China allotted itself more time than developed countries to rein in its emissions. Island nations and other vulnerable parties won mention of 1.5°C in the warming target. All of this allowed parties to leave Paris on a celebratory note, calling it "a tremendous collective achievement" (European Union), "a marvelous act" (China), "a resounding triumph of multilateralism" (St. Lucia), and "a tremendous victory for the planet" (United States).[43]

Given the praise heaped on the Paris Agreement, it is tempting to wonder whether its "how" and "who" of bottom-up pledges by all nations could have prevailed sooner. But even in hindsight, it may have been inevitable that top-down Kyoto-style mandates would have to fail before bottom-up pledges could be pursued as a plan B. History's most successful environmental treaty, the Montreal Protocol, had tackled ozone depletion with a top-down, developed-countries-first approach. As Kimble told me, "It's common for people solving big problems to look at solutions that have worked." Negotiators in Rio and Kyoto had seen the Montreal Protocol as the solution that was working for the other grand challenge of the global atmosphere, ozone depletion. Countries had not demonstrated that they would act on voluntary pledges; in fact, they still have not, as many Paris pledges remain unfulfilled.[44]

In any case, the Paris Agreement, updated by COP talks each year, is now the defining document for international climate policy. Countries approved the agreement so quickly that it officially entered into force on November 4, 2016. That was four years ahead of schedule, but also four days before the election of the president who subsequently pulled the United States out of the agreement. Still, by July 2020, the Paris Agreement had been signed by virtually all of the 197 parties to the UNFCCC and ratified by 189 of them. Only the United States temporarily withdrew, and after the election of President Joe Biden it soon rejoined the agreement. Unlike the skimpy Copenhagen Accord, the Paris Agreement weighs in at a hefty 29 articles after a lengthy preamble. Rather than slogging through all of the articles, I'll focus here on the landmark goals set by Articles 2 and 4.[45]

A Measure of Degrees

Article 2 of the Paris Agreement states the accord's most heralded aim: "Holding the increase in the global average temperature to well below 2°C above pre-industrial levels and pursuing efforts to limit the temperature increase to 1.5°C above pre-industrial levels." In the public imagination, this temperature target has become synonymous with the accord itself. An Eiffel Tower emblazoned with "1.5 DEGREES" welcomed delegates to Paris; "1.5 to well below 2 degrees" has become shorthand for the agreement ever since.

Iconic as the temperature target has become, it wasn't a given that climate diplomacy would be framed in terms of a warming limit. After all, the Kyoto Protocol, like the Montreal Protocol, had been framed in terms of timetables and emissions limits instead. The concept of warming limits emerged somewhat accidentally from the writings of Yale University economist William Nordhaus, who in 1975 mused that "it seems reasonable to argue that the climatic effects of carbon dioxide should be kept well within the normal range of long-term climatic variation." Nordhaus thought that this range topped out at 2 or 3 degrees above temperatures at the time. More recent analyses suggest that temperatures last peaked around 12,000 years ago at just 1 to 2 degrees above pre-industrial levels, closer to the Paris thresholds.[46]

Nordhaus, who in 2018 won the Nobel Prize in Economics for integrating climate change into economic analysis, never intended for his thought experiment about a warming limit to become the basis for climate policy. In fact, he considered any warming limit to be "deeply unsatisfactory," preferring that society aim to maximize benefits minus costs. Other scholars have criticized temperature targets as a way for "politicians to pretend that they are organizing for action when, in fact, most have done little."[47]

Nevertheless, warming limits have taken hold because they ground climate diplomacy in terms of the risks we are trying to avert and allow outcomes to be readily tracked. By contrast, aiming to optimize benefits minus costs would rely on wildly uncertain economic predictions. Valuations of the benefits of curbing climate change vary by an order of magnitude, since it is impossible to predict precisely how human health and the economy will be affected. For costs, industry representatives and regulators have historically vastly overpredicted the costs of controlling emissions of other air pollutants. They often fail to foresee innovations that arise once industries respond to incentives or mandates, as regulation drives innovation. As Chapters 4–7 show, emerging technologies could enable us to transition away from fossil fuels and remove carbon dioxide far more affordably than was once thought.[48]

Given that most of the Paris delegates agreed that a warming limit should be set, the levels set in Article 2 reflected a compromise among Saudi Arabia, which opposed any mention of 1.5°C; wealthy nations, which backed a 2°C target; and the Alliance of Small Island States and other "climate vulnerable" nations, which saw 1.5°C as a threshold for ensuring their survival. Although the "linguistic gymnastics" of Article 2 achieved the consensus required under UNFCCC protocol, it leaves the temperature target open to wide differences in interpretation. Since we have already warmed by roughly 1.2°C, a limit of 2°C allows for more than twice as much future warming, and thus more than twice as much emissions, as a 1.5°C limit.[49]

Aiming for Zero

Ambiguities in its phrasing and scientific uncertainties in relating it to emissions budgets make Article 2's temperature target a shaky foundation for policy. The Paris Agreement's more actionable target comes in Article 4: "Parties aim to reach global peaking of greenhouse gas emissions as soon as possible, recognizing that peaking will take longer for developing country Parties, and to undertake rapid reductions thereafter . . . so as to achieve a balance between anthropogenic emissions by sources and removals by sinks of greenhouse gases in the second half of this century, on the basis of equity, and in the context of sustainable development and efforts to eradicate poverty." Climate policy scholar Oliver Geden describes Article 4 as setting countries on a "race to net zero," with mere decades to reach the finish line. Unlike trying to squeeze emissions into a carbon budget, when the goal is net zero, every source must be balanced by a sink. As Geden told me, "When you frame things as a path toward zero, building a coal power plant becomes so blatantly inconsistent with your goal."[50]

Still, the phrasing of Article 4 contains crucial ambiguities. The English, Chinese, Russian, and Arabic versions leave unclear whether "anthropogenic" (human-caused) applies to sinks as well as sources, but the equally authoritative French and Spanish versions indicate that it does. Anthropogenic sources such as power plants and vehicles are relatively well-defined in UNFCCC reporting guidelines, but anthropogenic sinks from agriculture and forestry are more open to interpretation, especially when managed lands are affected by natural processes. Furthermore, not all sinks are created equal. Carbon sequestered deep underground can remain there for millennia, but carbon absorbed by forests could be re-emitted by the next wildfire or pest infestation.[51]

"Greenhouse gases" were made the gauge of the net-zero balance at the insistence of fossil fuel exporting countries, which did not want attention focused solely on carbon dioxide emissions. Ironically, net-zero greenhouse gases ("climate neutrality") is actually tougher to reach than net zero for carbon dioxide alone ("carbon neutrality"). This is true because some greenhouse gases like nitrous oxide lack anthropogenic sinks, so balancing their emissions actually requires net-negative carbon dioxide, with sinks exceeding emissions. The IPCC has found that holding warming below 1.5°C will likely require carbon neutrality by around 2050 and climate neutrality before 2070.[52]

Referring to greenhouse gases collectively also opens up questions about how to value gases with very different lifetimes in the atmosphere. Traditionally, this is done by comparing the heat trapped by each gas over a 100-year time span. But calculations of this metric are uncertain. Short-lived gases such as methane will cause a lot more warming now, and a lot less later, than a 100-year metric would suggest. Greenhouse gases also differ in their side effects, such as the ground-level ozone smog that forms from methane, the stratospheric ozone depletion caused by CFCs, and the ocean acidification caused by carbon dioxide.[53]

The "second half of this century" opens up a fifty-year window for achieving a net-zero balance. Warming could vary by a degree or more depending on when we reach net zero and the path we take to get there. Article 4 specifies that developing countries can take longer before their emissions peak, but it leaves unclear how to determine "equity," "differentiated responsibilities," and "respective capabilities." Most scholars agree that the United States, given its outsized responsibility for historical emissions and its relative wealth of financial and natural resources to pursue clean energy and carbon sinks, should be among the earliest nations to achieve net zero. "There's a need for major emitters to move more quickly so we have confidence that we will meet the global average time frame," said Kelly Levin, who has written extensively about climate targets for the World Resources Institute. "The United States and ideally a few other major emitters beating the global average gives us a higher chance of actually reaching those global average goals at the right time." Of course, under the Paris approach, each nation chooses the timing and scope of its own commitments and how vigorously to pursue them. Given the ambiguities of Articles 2 and 4, it is those national choices in the years ahead that will determine the future of our climate.[54]

National Contributions

While Articles 2 and 4 of the Paris Agreement set the goals, the workhorses for achieving these targets are the nationally determined contributions. Unfortunately, the intended contributions that countries submitted by 2020 are nowhere near sufficient to meet the agreement's twin goals. Climate Action Tracker, a consortium of three research organizations, estimates that even if every nation fully achieves its pledge, temperatures will rise nearly 3°C above pre-industrial levels by 2100 and continue rising thereafter. Since most nations are off track from achieving their pledges, actual warming will likely be slightly higher. As climate policy scholar David Victor of the University of California at San Diego told me, "We've got an oversupply of ambitious targets and goals, and an undersupply of action."[55]

The United States committed to reducing net greenhouse gas emissions 26–28 percent below 2005 levels by 2025 and reiterated

President Obama's ambition for an 80 percent reduction by 2050. Its Paris pledge did not set a timeline for net zero, although a subsequent White House strategy document in 2016 showed that cutting overall greenhouse gases by 80 percent would actually entail cutting carbon dioxide by nearly 95 percent, nearing carbon neutrality. President Joe Biden in 2021 announced his intention to strengthen the U.S. target to a 50–52 percent reduction by 2030 and net zero by 2050.[56]

The United States is far off track from achieving those targets. Net emissions fell 10 percent from 2005 to 2018, but much of that progress came from special circumstances: the recession that suppressed energy demand from 2007 to 2009; a subsequent recovery package signed by President Obama that subsidized energy efficiency and clean energy; a shale boom that helped natural gas outcompete dirtier coal but risks crowding out renewable alternatives; and temporary tax credits for energy efficiency, wind power, solar power, and electric cars. The COVID-19 pandemic temporarily curbed emissions, mostly due to short-term dips in vehicle use and air travel, but its long-term impacts remain unclear. In 2021, the Energy Information Administration projected that in the absence of new policies, energy sector emissions will stay flat through the middle of the century, putting the U.S. emission reduction targets out of reach. Climate Action Tracker expects most other countries to miss their targets too.[57]

The Road Ahead

The Paris Agreement ended the futile quest for setting top-down national greenhouse gas limits at a global negotiating table, as the Montreal Protocol had done for niche CFCs. Carbon dioxide is too pervasive a gas, and the U.S. Senate too wary a body, for even weak caps to be set at faraway conferences. The Kyoto Protocol tried to do just that and left climate diplomacy adrift for decades. The Paris Agreement provided a fresh start by letting countries set their own bottom-up commitments.

The Paris Agreement can be only as successful as the commitments themselves. So far, the pledges are no match for the treaty's aspirations, and countries face little consequence for setting them too weak or failing to uphold them. That makes it easy to dismiss the treaty as yet another failure of climate diplomacy.

With the climate warming so fast, it's too late to start from scratch and build global consensus for a new approach. The Paris Agreement is what we've got in the global arena. Fortunately, less heralded features of the treaty beyond its headline targets provide tools for strengthening the treaty from within, as I'll explain in the next chapter.

The international key to confronting climate gridlock won't be seized by global consensus-driven approaches alone. More targeted and nimble alliances of the willing must augment the global framework. As Susan Biniaz has written, "the [Paris] Agreement cannot meet the enormity of the challenge on its own." Emerging ideas for climate clubs and other measures offer opportunities for reinforcing and acting beyond the Paris Agreement, as the next chapter will reveal.[58]

three THE ROAD FROM PARIS

The Paris Agreement's requirement for each country to set its own national commitment poses a dilemma. The more ambitious the commitment, the costlier it may be to attain, at least if clean energy is costlier than fossil fuels. However, if aggregate commitments are weak, warming will continue unabated. A long-ago experiment with a classic game sheds light on how countries should handle this dilemma. Looking ahead, innovative diplomacy within and beyond the Paris framework along with emerging technologies could upend the dilemma itself by creating a self-interest in greater ambition.

Insights from a Classic Game

Invitations were mailed from University of Michigan political scientist Robert Axelrod recruiting competitors for a friendly tournament. It was an eclectic invitation list: mathematicians from Texas, New York, and Switzerland; economists from Yale and Virginia; psychologists from Canada and North Carolina. Across their disparate disciplines, the invitees shared one common trait: an expertise in game theory about the game they were asked to play, the prisoner's dilemma.[1]

The prisoner's dilemma mimics the choices faced by two prisoners who conspired in a crime. The prisoners sit in isolated cells as they are interrogated by prosecutors. If both prisoners cooperate with each other and stay silent, both can escape conviction. If only one prisoner defects, he can testify that the other committed the crime and be rewarded for his testimony. If both prisoners defect, both will be convicted and share responsibility for the crime. The crucial feature of the game is that the conspirators are collectively better off if they cooperate with each other in staying silent. But individually, not knowing his conspirator's response, each prisoner is personally better off defecting.

Fourteen of the game theory experts accepted Axelrod's invitation. Each submitted a program instructing the computer on a strategy for what moves to make. The year being 1979, the programs were written in FORTRAN or BASIC and mailed in on paper. Axelrod and his assistant coded them into a computer and paired each entry with each other in 200-round matchups. Although the players could not communicate directly, the 200-round format meant that their strategies could account for each other's previous moves when setting their next move.[2]

Tallying the scores, Axelrod was stunned to see the results. The shortest entry, a mere four lines of FORTRAN submitted by mathematical psychologist Anatol Rapoport, had won the tournament. Axelrod then posted ads in computer journals inviting entries for a second tournament. He provided the strategies and results from the first tournament for contestants to consider as they crafted their strategies. This time, entries poured in from sixty-two players from six countries, ranging from a ten-year-old computer hobbyist to the same game theorists who had competed in the first round. But none of them managed to beat Rapoport, who won again with the same four lines of code.[3]

Rapoport's winning strategy was simple: tit for tat. Cooperate in the first round. Then, in each other round, do whatever the other player did in the previous round. In other words, retaliate once after each defection and cooperate after each cooperation.

Why did this simple tit-for-tat strategy succeed while others floundered? Axelrod devoted an entire book, *The Evolution of Cooperation*, to drawing lessons from his tournaments and related game theory. His purpose was to address a central question that implicitly encapsulates the challenge of climate diplomacy: "Under what conditions will cooperation emerge in a world of egoists without central authority?" Axelrod concluded that cooperation could indeed emerge as a winning strategy, even without a central authority to enforce cooperation on selfish players, if the playing field was aligned properly and players adopted wise strategies.[4]

In particular, Axelrod noted that four traits set apart the high-scoring strategies in his tournaments: they were nice, retaliatory, forgiving, and clear. Being "nice" by not defecting first mattered most of all. That is a paradoxical result, since the prisoner's dilemma rewards each player for defecting, if the other player's moves are a given. But in the tournaments, as in life, the actions of others are not a given. Our history of interactions, whether in daily encounters or climate action, can influence each other's next steps. Niceness can succeed if it elicits niceness from others.[5]

Niceness alone won't elicit niceness. Instead, it invites exploitation. Rapoport's strategy avoided that fate by starting out nice, retaliating against meanness, and then being forgiving—returning to niceness as soon as the other player did too. Moreover, his strategy was clear, so others soon learned that their niceness would be reciprocated.

Rapoport's victory came not by domination but by fostering mutual success. In fact, he did not beat the other players in head-to-head matchups. Instead, his strategy maximized his cumulative scores, and thus won the tournaments, by eliciting more cooperation than any other, enabling both players to score highly in his matchups. Blending niceness, retaliation, forgiveness, and clarity bred a virtuous cycle of cooperation, even without a higher authority to enforce it. Those four simple traits helped everyone score higher, fostering mutual success as will be needed to confront climate change.

Lessons from Game Theory for Climate Diplomacy

Axelrod's book discussed how the four winning traits—being nice, retaliatory, forgiving, and clear—could help cooperation emerge to address an array of challenges: nuclear arms control, divorce settlements, business competition, congressional deal-making, and so on. Not mentioned among those challenges was climate change. After all, the book was published in 1984, eight years before the Rio summit. Dozens of books and hundreds of articles have been written on the game theory of climate change since then, but the most crucial lessons still trace back to those four winning traits. Although scholars are increasingly recognizing how climate diplomacy differs from the prisoner's dilemma, Axelrod's quartet of winning traits remains as central as ever to successful outcomes.[6]

Mitigating climate change is a particularly vexing challenge, in part because of the mismatch between who benefits and who pays. The benefits of a stable climate, though vast, are uncertain and dispersed across billions of people, many living in vulnerable countries with little geopolitical power and scant emissions to control. The costs of mitigation, by contrast, are borne mostly by polluting industries that may lobby aggressively to oppose costly actions. Moreover, policies whose impacts transcend centuries are being negotiated by politicians focused on the next election, with no input from yet-to-be-born generations. All of this creates a self-interest to do less, even if the world would be better off if everyone did more. Some actions will be pursued because they yield immediate local benefits, such as energy savings, jobs, and cleaner air and water. However, deeper cuts in emissions beyond local self-interest will require cooperation to emerge among selfish nations.

Key features of Axelrod's tournaments hold true in climate diplomacy. Parties focus mainly on their own self-interest. They know their shared history, but not their future. Cooperation is mutually beneficial but may be individually costly. There is no central authority to impose a deal or enforce compliance.

Distinctions between the prisoner's dilemma and climate diplomacy must be noted too. Unlike Axelrod's contestants, world leaders and diplomats can and do communicate. Each country's actions are determined not by a single leader or algorithm, but by the interplay of numerous actors at all levels of government, industry, and society. Many influential actors in government and industry, along with the majority of voters in domestic and global surveys, increasingly want their country to act on climate regardless of what other countries do. Their desires reflect rising concern about climate change and a growing awareness that clean energy yields local benefits beyond the global benefits to climate. But they must overcome inertia and face pushback from industries and politicians opposed to climate action. To help them do so, political scientists Michaël Aklin and Matto Mildenberger have written that climate diplomacy "should focus on empowering pro-climate constituencies" to seek stronger domestic actions, rather than on preventing the free-riding that is central to the prisoner's dilemma. Nonetheless, the same winning traits that fostered cooperation and deterred free-riding in Axelrod's tournaments can help provide that empowerment.[7]

A History of Flawed Strategies

Reflecting on the history of climate diplomacy, we see that each major player has deviated from Rapoport's four winning traits: niceness, retaliation, forgiveness, and clarity.[8]

The United States has been neither nice nor clear. It was the first nation to defect from the Kyoto Protocol and the only one to leave the Paris Agreement. Its contributions to the Green Climate Fund and other foreign aid have been paltry. Clarity vanishes with each election. "Every four or eight years, each party undoes everything the other party did," said environmental policy scholar Benjamin Sovacool, who has tallied more energy policy reversals in the United States than in any other developed country.[9]

However, when the United States does act, it usually insists on reciprocity from other countries. For example, the Montreal Protocol extended and internationalized domestic CFC restrictions that the United States had already adopted. The Senate's Byrd-Hagel resolution insisted that any treaty limiting U.S. emissions must limit developing countries' emissions too. Most American bills and proposals for carbon pricing have included provisions to impose a carbon tariff or "border adjustment tax" on imports from any country that does not cap or tax carbon as well. Such tariffs can reassure American industries and workers that carbon pricing will not place them at a competitive disadvantage, and they may prod other countries to establish their own carbon pricing. A border adjustment tax was a prominent feature of the Waxman-Markey cap-and-trade bill that passed the House but not the Senate in 2009; the carbon fee-and-dividend plan crafted by retired Republican statesmen in 2017; the Green New Deal proposed by liberal Democrats in 2019; and the campaign platforms of Joe Biden and other Democratic presidential candidates in 2020. Of course, none of those proposals became law, and emissions remain unpriced in most of the United States outside of California and power plants in northeastern states. Nevertheless, this tradition of reciprocity being demanded by the largest economy in the world provides enormous leverage to internationalize any domestic actions that are taken. Trouble is, the United States hasn't been nice enough to take many actions that are worth spreading.[10]

The European Union has been far nicer, seeking cooperation at every turn and accepting some of the steepest emissions cuts in the Kyoto and Paris agreements. Its emissions trading system and other policies have yielded substantial progress. But Europe has been less concerned with reciprocity than the United States, choosing not to retaliate against inaction by the United States and others. When the United States failed to ratify the Kyoto Protocol and some other countries withdrew or missed their targets, Europe imposed no consequences and remained committed to its target.

While the European Union has been pricing carbon through cap-and-trade for its own industries since 2005, it has not coupled domestic pricing with a carbon tariff as American legislation would have done. Ahead of the Copenhagen talks in 2009, French president Nicolas Sarkozy urged the European Union to impose a carbon tariff on imports from countries that did not price carbon. But in retaliation-averse Europe, Germany's environment secretary denounced Sarkozy's proposal as "eco-imperialism." Sweden's environment minister called it "green protectionism." Sarkozy revived the idea of carbon tariffs while campaigning to retake office in November 2016, calling for them to be imposed on imports from the United States if president-elect Donald Trump carried through on his threats to quit the Paris Agreement. The idea was again rejected by European leaders. When President Trump announced in 2017 his intention to withdraw from the Paris Agreement, Europe again imposed no consequences and instead reiterated its commitments. China has allowed emissions to soar with little consequence from Europe.[11]

The European Green Deal passed by the European Commission in 2020 unilaterally strengthened the EU's Paris pledge for 2030 by an extra 10 to 15 percentage points and committed the bloc to becoming climate neutral by 2050. That may have given Europeans the moral high ground for their niceness, but it reiterated to other countries that there would be no retaliation for their inaction. Why should other countries be nice when they can free-ride on the niceness of the Europeans?[12]

An insistence on reciprocity may finally be emerging at the time of this writing. In the December 2020 Brexit deal, the EU insisted that the United Kingdom reaffirm its aims for climate neutrality by 2050. The EU was developing plans to impose a carbon tariff in 2021.[13]

Rapidly growing countries such as China and India have not always been nice or clear. Their refusal to accept emissions targets in the Kyoto Protocol undermined its prospects in the U.S. Senate. Many developing countries have shifted in and out of alliances with unclear bargaining positions. Meanwhile, island states and other vulnerable nations have at times been unforgiving, seeking reparations for damages from past emissions. Such demands may seek to address injustices but can stifle cooperation on mitigating future harm. China's pledge in 2020 to go carbon neutral by 2060 could be a major turning point. If fulfilled, that commitment would cut projected warming by 0.2 to 0.3°C and set an example for other growing economies to follow.[14]

There are self-interested advantages to not being nice, retaliatory, forgiving, and clear. Not being nice can be profitable if it triggers no retaliation from others. American industries have prospered by burning cheap fossil fuels without paying carbon taxes. China industrialized its economy with cheap coal while resisting emissions limits in treaties. Not retaliating against the United States and China makes sense for countries that fear triggering a trade war with the world's largest economies. Public support for climate action in Europe has been strong, even when actions are taken unilaterally. Forgiveness is a bitter pill to swallow for island nations that see their very existence threatened by sea level rise. Clarity is difficult to maintain in the United States, where political power swings with each election and a minority of senators can block legislation and treaties. Yet whatever the reasons, the flawed strategies of the major players have left us with what climate scholar David Victor called, in the title of his book, global warming gridlock. Adopting the four winning traits that Axelrod identified for eliciting cooperation will be key to confronting that gridlock. I'll turn first to how cooperation can emerge in the global arena under the Paris Agreement before exploring what should be pursued in more targeted partnerships beyond it.[15]

Opportunities Within the Paris Framework

The framers of the Paris Agreement crafted it more as a constitution than a detailed code of law. They did so by choice, to enable details to evolve in an overarching framework, and by necessity, to paper over differences with clever ambiguity and win acceptance from nearly two hundred countries. At this point, it's too late to replace the Paris Agreement with an entirely new framework. Emissions reductions must begin now. Renegotiating a global framework would take years. Thus, what we need is to work more effectively within the Paris framework, while also pursuing other actions beyond it. "We don't need to negotiate a new agreement; all we need to do is make Paris work," said Melinda Kimble, the former U.S. climate negotiator.[16]

Making Paris work will depend on reinforcing the agreement's three main components: the rulebook, stocktakes, and nationally determined contributions or "commitments."

The rulebook fleshes out how the Paris Agreement operates and how countries report progress on their national commitments. Delegates established an initial 133-page rulebook at the 2018 climate summit in Poland. Contentious follow-up talks in Spain in 2019 yielded little headway, and the 2020 summit was postponed due to the COVID pandemic.[17]

Operating under the guidance of the rulebook, the five-year cycles of conducting stocktakes and updating national commitments are the "beating heart" of the Paris Agreement. Since the initial Paris commitments were widely seen as inadequate for meeting the agreement's temperature targets, it is these cycles that offer hope of bending emissions curves toward achieving them.[18]

From a game theory perspective, the cycles turn climate diplomacy into an "iterated" (repeated) game. If a game like the prisoner's dilemma is played just once, non-cooperation is the winning strategy. Cooperation can emerge if players use their actions in each round to signal what their counterparts should do in subsequent rounds. That's what Rapoport's winning tit-for-tat strategy did so effectively. Axelrod called it the "shadow of the future"—the more that players expect to be interacting in the future, the greater their incentive to cooperate today. Similarly, knowing that more rounds of stocktakes and updates lie ahead creates accountability, spurring countries to uphold and enhance their commitments and enabling them to set consequences for others that fail to do the same. A virtuous cycle of accelerated action can take hold. Cooperation can indeed emerge.[19]

Stocktakes were established as a means of assessing progress ahead of each round of updated commitments, which are due two years later. The first stocktake, scheduled for 2023, will set crucial precedents for future ones. The United States could be pivotal to seeking rigor in the stocktakes. As William Sweet wrote in his history of climate diplomacy, the United States consistently advocated transparent and rigorous accounting and reporting, even when its commitment to action wavered. It was American negotiators who insisted that the Paris Agreement feature binding provisions for reporting, even though emissions commitments are non-binding. The hope was that transparent reporting and peer pressure would enable countries to prod each other to continue ratcheting up their efforts. Even at the Trump-era climate summits in Poland and Spain, "transparency remained a top priority for U.S. negotiators," said Simon Evans, who covers the summits as deputy editor of *Carbon Brief.* Trump's team continued to ally with European negotiators to insist on transparent reporting standards in the rulebook. "With his administration apparently only caring about the 'we're out of Paris' headlines, the word at the [summits] was that the U.S. negotiating team continued much as before, . . . negotiating with an eye on U.S. reentry," Evans told me.[20]

Stocktakes will not only inventory emissions but also assess the effectiveness of control measures and efforts for promoting adaptation and resiliency, financial support for developing countries, and technology development and transfer. At the insistence of vulnerable nations at the Poland summit, loss and damage caused by climate change were added to the scope. The vast breadth of the stocktakes will make them unprecedented opportunities to gauge what's working and what gaps remain. Done well, the stocktakes will yield an unparalleled array of data to inform future efforts. The collaborations across governments, industry, and academia that will be needed for rigorous stocktakes could form the basis for ongoing reviews of progress. Done poorly, the stock-takes could undermine hopes for clarity and accountability.[21]

The 2023 stocktake is likely to portray a yawning gap between ambition and reality. Contributions to the Green Climate Fund to support adaptation and emissions mitigation in developing countries cover only a tiny fraction of the $100 billion per year intended to come from wealthier countries. Emissions have continued to rise, rather than falling toward a net-zero balance. Even if all countries fully achieve their commitments issued by 2020, that would merely be enough to flatten global emissions rather than reduce them by roughly 25 percent by 2030 as needed to be on pace for a 2°C limit, or 50 percent as needed for 1.5°C, according to the United Nations Environment Programme. If countries remain off track from meeting their commitments, actual gaps could grow even larger. All this leaves the world on pace for roughly 3°C of warming above pre-industrial levels by 2100, unless China and other countries begin backing up their recently announced net-zero aspirations with firmer actions.[22]

Following each stocktake, countries must update their national commitments. Rapoport's four winning traits can inform those updates. A vigorous rulebook can ensure the clarity of commitments. If Europe and other historically "nice" players act with more reciprocity and willingness to retaliate, they could prod laggards like the United States and China to do more. If the United States and China play nice by backing up ambitious aspirations with vigorous domestic policies, they could send a powerful signal for other countries to follow suit. The United States has especially strong leverage, given its tendency to demand reciprocity. Emerging from the Trump era, when the United States stood alone in shirking the Paris Agreement, a little bit of niceness could go a long way.

Clarity is also crucial to climate diplomacy. Globetrotting diplomacy by the French ahead of the Paris talks forged a spirit of compromise, which had been elusive in Copenhagen. The United States and China prodded each other to pursue more ambitious commitments, as announced by Presidents Barack Obama and Xi Jinping at their 2014

Beijing summit. Unfortunately, too many other nations developed their initial commitments in relative isolation.[23]

More extensive bilateral and multilateral diplomacy is needed for countries to propel each other toward greater ambition in each cycle of stocktakes and enhanced commitments. Coordinating commitments with trade rivals and allies alike can give countries confidence that they will not put their industries at a competitive disadvantage. Further confidence and leverage can be gained by making pledges conditional on the actions of others. "A pledge should be set up as 'I'll do X, but I'll do X+2 if other countries do X+2,' " suggests Victor. Many developing countries set their Paris commitments on a conditional basis. For example, Mexico pledged unconditionally to reduce emissions by 25 percent below business as usual by 2030, but by 40 percent if there is a global agreement for carbon pricing and technical support. Morocco pledged to reduce emissions by an additional 25 percent by 2030, conditional on receiving financial support. So far, those conditions are not being met. Globally, neglecting the conditional pledges would forgo the opportunity for 3 billion tons of annual carbon dioxide emissions reductions by 2030.[24]

History shows that for the United States, domestic policy advances should precede international commitments, not the other way around. The Montreal Protocol built upon CFC restrictions that were already underway in the United States. The Kyoto Protocol, by contrast, set an emissions cap for the United States that was not backed by domestic policy. As former U.S. climate negotiator Susan Biniaz recalls, "We had very little leverage or diplomatic ability to go to other countries to ask them to take on measures or targets when we had no domestic measures or targets ourselves."[25]

Embedding domestic policies as updates to Paris commitments could make them more durable. If other countries adopt policies contingent on those commitments, reneging on them could trigger retaliation. Reciprocity by allies and rivals might not trump the whims of a rash president or prime minister, but it would show persuadable legislators that there are consequences to abrogating a country's responsibilities. All of this depends on other countries adopting more of Rapoport's winning traits like clarity and retaliation, rather than leaving their strategies unclear or turning the other cheek to a country's retrenchment.

Specific policies enacted into law are the workhorses of Paris commitments. However, emissions targets and timetables provide valuable framing, even if policies have not yet risen to their ambition. When the Trump administration began maneuvering to exit the Paris Agreement, hundreds of businesses, states, and cities signed on to the America's Pledge and We Are Still In initiatives, in which they pledged to reduce emissions in line with the targets set in the U.S. commitment. Empowering pro-climate constituencies like these is one of the most important benefits of a climate treaty, and targets give them a focal point to align their efforts. When nations strengthen their targets and timetables, corporate and sub-national actors can strengthen their pledges accordingly.[26]

Opportunities Beyond the United Nations Framework

The United Nations–backed framework that was established in Rio and extended by the Paris Agreement operates on the basis of consensus among all parties. With nearly two hundred parties on board, the framework now involves nearly every nation on Earth.

Global consensus has its appeal. A global challenge would seem to require global solutions. Anything less would not tackle all emissions and would let outsiders free-ride on the contributions of members. Consensus is needed to avoid encroaching on the sovereignty of nations.

But a global consensus can only take us so far. Political scientist Arild Underdal warned that "collective action will be limited to those measures acceptable to the least enthusiastic party." At climate summits, the least enthusiastic parties are typically oil exporting countries such as Saudi Arabia. The wording of the Rio framework and the Paris Agreement had to be weakened to win their consent. The Kyoto Protocol set weak caps for rich countries and no caps for poor ones, yet still could not bring all parties to consensus.[27]

One way to sidestep the veto of the least enthusiastic countries in global forums is to create coalitions of enthusiastic ones in smaller settings. The simplest coalition is of course an agreement between two nations, forged by bilateral diplomacy. For example, Norway has partnered with Indonesia to reduce deforestation and fires there. Those fires are exceptionally polluting, since many Indonesian forests sit atop layers of peat that can smolder for weeks. Especially severe peatland fires in 2015 blanketed Southeast Asia with haze and led to more than 100,000 premature deaths. Fires can be averted relatively affordably, but Indonesia lacks the resources to do so on its own. The cleaner air, preserved forests, and assistance that come with this partnership provide Indonesians with ample reason to embrace it, even if it means forgoing some slash-and-burn agriculture and grazing. Meanwhile, Norway is already among the world's leaders in adopting clean energy and electric cars, and its population is small. Thus, its partnership with Indonesia expands its capacity to reduce emissions. The United States could pursue deals with allies in the Americas to protect forests, reduce livestock emissions, or store carbon in soils.[28]

Climate Clubs

Broader coalitions of ambitious nations could pursue greater progress by forming climate clubs. Such clubs could take a variety of forms, and a panoply of clubs could be more effective than any one club alone.

The most highly cited concept for a climate club arose in a 2015 paper by Yale University economist William Nordhaus, who made it the centerpiece of his acceptance speech for the Nobel Prize. Nordhaus suggested that club members should agree upon a carbon tax rate and impose carbon tariffs on imports from non-member countries that do not price carbon, much like the border adjustment taxes that have been staples of U.S. carbon pricing proposals.[29]

Nordhaus's approach is well grounded in economics, since carbon taxes promote cost-effective emissions reductions and a club would extend its reach. The accompanying carbon tariff would give other countries an incentive to join the club, so that their industries pay carbon taxes domestically rather than carbon tariffs to other countries. However, a tax-based club flies in the face of history and politics. The United States has never accepted climate mandates set in the international arena. Can we really expect that a club of countries will agree to a uniform carbon tax and maintain it through ever-shifting political and economic conditions? Perhaps a handful of countries would, but it's hard to imagine the United States doing so. Countries outside the club might retaliate with tariffs of their own. Moreover, a club of wealthy countries imposing carbon tariffs on imports from poorer ones would exacerbate inequality. "The idea of climate clubs deserves serious consideration, but it must actively guard against creating a new global governance regime that reinforces the economic and geopolitical imperialism of the past," said Arvind Ravikumar, an assistant professor of geosystems engineering at the University of Texas at Austin.[30]

A more flexible vision for climate clubs was introduced in 2011 by David Victor in his book *Global Warming Gridlock*. In Victor's approach, countries would come together much like they do for trade deals such as the North American Free Trade Agreement or the founding of the World Trade Organization (WTO) and set preliminary rules and expectations. Crucially, club members would convey exclusive benefits upon each other, such as favored trading status or reduced sanctions. Much like joining the WTO, each country would need to submit an accession deal committing to specific actions to join the club. Victor tells me that such a club would not replace the Paris framework but rather supplement it. In fact, policies pledged for a club should be embedded into a country's Paris commitment and assessed via the stocktakes and rulebook.[31]

Other climate clubs could coordinate clean energy research and share technologies that emerge. The International Solar Alliance has already been formed to accelerate the financing, development, and deployment of solar technologies. Similar alliances could be formed for offshore wind, geothermal energy, building efficiency, low-carbon materials, and carbon sequestration. Such efforts would create a "technology push" toward better and more affordable technologies. Countries could accompany that push with a "market demand pull" if they set standards or incentives to drive deployment of those technologies. As supply chains become increasingly global, manufacturers may opt to follow the strongest standards globally.[32]

For all of these clubs, setting exclusive benefits of membership can entice countries to join and discourage them from leaving. That way, membership can be sustained or expanded even as political leaders come and go. The more countries that join, the more valuable those benefits would become, creating self-reinforcing momentum. "As the club grows, then you have this snowball effect that would make the club more and more attractive," said political scientist Jon Hovi of the University of Oslo. In Nordhaus's tax-based vision, that snowball effect would come from the carbon tariffs, which would grow more important to avoid as more countries join the club and impose the tariffs on non-members. In Victor's trade-oriented vision, preferential trading status would incentivize membership. In technology-based clubs, access to new technologies could be made an exclusive perk of membership to attract and retain members. However, sharing technologies more broadly, at least to developing countries, would accelerate diffusion and make decarbonization more affordable worldwide.[33]

Different climate clubs would attract different members. For example, a Nordhaus-style club coordinating carbon pricing should consist only of those countries that can commit to enacting and maintaining such policies. A club focused on technology should include countries that are leaders in clean energy research and development but should also make technologies available to poorer countries that will need affordable options to decarbonize. Countries could join any number of these clubs with intertwined aims. Clubs could set their own evolving rules for entry and exit and ratcheting up ambition. Clubs could even design ways to partner with cities, states, or businesses. That could be especially important if gaps among local, corporate, and national ambitions widen as national politics swing.

International cooperation will also be needed to tackle emissions from key sectors. For example, the International Civil Aviation Organization has already developed carbon emissions standards for aircraft. The International Maritime Organization toughened its standards for sulfur emissions in 2020 but should do more to limit greenhouse gases. Nations in climate clubs could reinforce these efforts by granting preferential access at their airports and ports to airplanes and ships that outperform the international standards.

Diplomacy aimed at protecting stratospheric ozone has yielded crucial co-benefits for climate. By phasing out ozone-destroying CFCs that are also potent greenhouse gases, the Montreal Protocol and its amendments have done more to cool the climate than any climate treaty. Unfortunately, some ozone-friendly substitutes for CFC refrigerants such as hydrofluorocarbons (HFCs) are potent greenhouse gases too. Negotiators meeting in Kigali in 2016 agreed to amend the Montreal Protocol to phase down climate-warming refrigerants by 80 percent. That could save the equivalent of ten years' worth of carbon dioxide emissions and avert up to 0.4°C of warming—enormous amounts in the context of climate change. "It is likely the single most important step we could take at this moment to limit the warming of our planet and limit the warming for generations to come," Secretary of State John Kerry said at the time. By 2020, 99 countries and the European Union had ratified the Kigali Amendment, bringing it into force. But the Trump administration never sent the Senate the treaty to be ratified, despite bipartisan support from senators, environmentalists, and industrial groups. Legislation passed at the end of 2020 finally committed the United States to curtailing HFCs but did not ratify the treaty.[34]

Across these international approaches, priority should be given to empowering developing countries to transition sustainably to clean energy. As wealthier countries rein in their emissions, most growth in emissions is expected to come from developing countries that lack the resources to invest in costly technologies. Sustainable industrialization will depend on their ability to leapfrog over traditional fossil-based routes directly to clean energy, much as they skipped over landline phones and adopted cell phones. Technology transfers and cooperation can make clean energy more affordable in developing countries without politically unpopular direct payments from wealthier ones. But free access to patents won't be enough if countries lack the expertise and resources to deploy them. Support from nongovernmental sources such as philanthropists, the World Bank, the Green Climate Fund, and the European Investment Bank can help drive those deployments. The conditional pledges already made by developing countries in their Paris commitments offer billions of tons' worth of opportunities to achieve more rapid emissions reductions, but only if wealthier countries and institutions step forward with support. "Lower-income countries will ratchet up their commitments if it's accompanied by significant funding and tech transfer from the developed world," said University of Toronto political scientist Jessica Green. Unfortunately, wealthy countries and private entities have fallen far short of the $100 billion per year promised in Copenhagen and Paris.[35]

Technology transfers and cooperation should not be seen as a one-way street. Growing markets for clean energy in populous developing countries can drive down the costs of technologies worldwide, thanks to economies of scale and learning curves. As knowledge, expertise, and research institutions grow, developing countries will increasingly become sources of clean energy innovations themselves. International efforts must be pursued in ways that foster cooperation, not dependence. To extend an old adage, developing countries must not just be given fish or taught to fish, but also empowered to develop better ways of fishing that the rest of the world can adopt.[36]

Supply-Side Options

Most of these ideas for complements to global climate treaties, like those treaties themselves, focus on reducing demand for fossil fuels and associated emissions. The supply of fossil fuels has received far less attention in climate diplomacy. As economist Geir Asheim of the University of Oslo told me, "There's an asymmetry in the way climate economics has confronted this problem," focusing far more on demand than supply.[37]

Activists have long focused on the supply side of the climate challenge. In fact, the most ardent protests outside of climate talks often call for tougher restrictions on fossil fuel supply. Coal mines, fracking sites, pipelines, and other elements of fossil fuel supply trigger more visceral opposition than invisible greenhouse gases, and they are more concentrated than the millions of vehicles and facilities that demand their fuels. As scholars increasingly consider "keep it in the ground" as more than just a protest slogan, they estimate that roughly 80 percent of coal, half of natural gas, and a third of oil reserves globally must ultimately be kept in the ground to hold warming below 2°C. More immediately, the United Nations Environment Programme has found that fossil fuel production is on track to be more than twice as high in

2030 as would be consistent with a 1.5°C target, and that 6 percent per year cuts in fossil fuel supplies would be needed to close that "production gap." For 2°C, 2 percent per year cuts are needed.[38]

Of course, in the end, supply roughly matches demand. Coal isn't piling up mined but unused. The difference between demand-side and supply-side approaches is more than semantics, however. Driving down demand for fossil fuels in only part of the world, as in a climate club or a single nation's Paris commitment, makes fossil fuels more affordable, allowing demand to rebound elsewhere. Lower prices also reduce the motivation to develop clean energy technologies. If owners of large fossil fuel reserves expect prices to continue falling, they could face a perverse incentive to extract them as soon as possible to avoid leaving those assets stranded. Economist Hans-Werner Sinn calls this the green paradox.[39]

By contrast, supply-side restrictions on the mining, drilling, and export of fossil fuels make them scarcer and costlier on global markets, thereby curbing demand. However, those higher prices could make it more profitable to produce more fossil fuels where restrictions are not in place. Minimizing a rebound in supply would require complementary policies to curb demand or assembling a supply-restricting coalition large enough to outmatch the rebound that could occur elsewhere.

A supply-side climate club centered around oil seems ill-advised. After all, the world already has a supply-side oil club—the Organization of the Petroleum Exporting Countries (OPEC). OPEC has struggled to manage oil markets, and it focuses on the priorities of exporters, not climate.

However, a supply-side climate club could succeed for coal. Coal makes up the largest share of carbon emissions today and of fossil fuels still in the ground. The United States would have substantial leverage in a club intended to constrain coal supply. It sits atop more recoverable coal reserves than any other country—230 billion tons, or 22 percent of the global total. Despite its mythic stature politically, coal represents a tiny share of the American economy and employment. Coal mines employ fewer workers than Arby's sandwich shops, so economic revitalization of mining communities is not an insurmountable task. The market value of the U.S. coal mining industry fell to $20 billion in 2020, roughly one-fourth the value of Snapchat.[40]

Restricting coal mining on public lands and banning coal export terminals could keep American coal from reaching foreign markets. But the United States need not go it alone. A key American ally, Australia, holds 14 percent of global coal reserves, yet the coal industry employs just a sliver of the Australian workforce. India and China together hold 22 percent of global coal reserves and are plagued by the smog from coal burning. The more coal-rich countries commit to keeping their coal in the ground, the costlier and scarcer coal will become.[41]

President Obama began to pursue some supply-side restraints in his second term. He rejected the Keystone XL oil pipeline by arguing that climate change requires keeping most fossil fuels in the ground. He also placed a moratorium on new leases for coal mining on federal lands and banned oil drilling in parts of the Arctic and Atlantic coasts. "The world needs a big fossil fuel producer to stand up and say we're going to phase down fossil fuel [supply] . . . and we're going to do this for climate reasons; Obama started going down that road," said Georgia Piggot, who studies supply-side policies at the Stockholm Environment Institute. Soon after taking office, President Joe Biden paused leases for oil and gas drilling on public lands and waters, restricted drilling in the Arctic, and reversed the Trump administration's approval of Keystone XL.[42]

Economists such as Asheim and Bård Harstad of the University of Oslo have suggested that supply-side measures should be incorporated into future national commitments under the Paris Agreement or international climate clubs. Doing so would be controversial. Victor told me it would be "bonkers" to think that enough producers would restrict supply to avert a rebound abroad. Asheim responded, "It's even more bonkers to think you solve this on the demand side alone," after three decades of demand-oriented climate treaties have failed to rein in emissions.[43]

Linkages Across Diplomacy, Technology, and Policy

Whether in Paris Agreement commitments or in smaller clubs, national policies remain the building blocks of global progress. National policy cannot be set in Rio or Kyoto or Paris. Bolder domestic action can only begin with policies enacted by national leaders. Those policy decisions will be shaped by the availability of affordable technologies and the influences of public opinion, industries, and organized groups. But once domestic policies are passed, embedding them in conditional Paris commitments, bilateral deals, and climate clubs can leverage them to impel others to act.

America's economic heft and its traditional insistence on reciprocity position it to be a driver of international climate action, if it can ever become nice and clear. Given America's historical lack of niceness and clarity, climate clubs and other international means of creating consequences for domestic actions will be crucial for keeping the United States on track as political winds shift. Climate-friendly leaders should embed domestic advances into international deals while they are in power, to sway the actions of other leaders who may follow. The Paris Agreement and emerging ideas for climate clubs provide more opportunities to do so than ever before, bringing the international key within reach of an American president and allies who choose to seize it.

Realigning the international playing field can be a powerful factor in motivating climate action domestically and leveraging it to spur action abroad. But the international key can't unlock climate gridlock on its own. Domestic policies will be needed to uphold international commitments and to have domestic advances worth leveraging. "Until we are able to prove we are able to address climate change constructively and aggressively domestically, our authority to sell our ideas internationally will be limited," Kimble told me. Meanwhile, better technologies will be needed to transition away from fossil fuels and sequester the emissions that remain, making action more affordable domestically and abroad.

We turn to those technologies next.

four PILLARS OF DECARBONIZATION

However successful diplomacy may be, decarbonization is at heart a series of national endeavors rooted in technology. Transitioning each sector of the economy from fossil fuels to clean energy and offsetting the emissions that remain will require extraordinary technological transformations. That includes rapidly deploying clean technologies that already outperform their rivals; driving down the costs of those that are not yet cost-competitive; and developing new technologies where viable options are not yet available.

With so much emissions to cut—roughly 6 billion tons per year in the United States alone and over 40 billion tons globally—it helps to begin with the adage famously misattributed to the bank robber Willie Sutton: "Go where the money is." Fossil fuels is where the emissions are. Burning them is the source of more than 90 percent of carbon dioxide emissions nationally and globally. Obtaining and transporting them results in leaks of methane from mines, drilling sites, and pipelines. All other emission sources, including carbon dioxide from cement, methane from livestock, and nitrous oxide from agricultural soils, are small relative to fossil fuel emissions.[1]

Fossil fuels in the United States are burned mainly for transportation (38 percent of U.S. fossil fuel carbon dioxide emissions in 2019), power plants (33 percent), industry (17 percent), homes (7 percent), and businesses (5 percent), according to the U.S. EPA. Globally, the shares of energy-related carbon dioxide from power plants (40 percent) and industry (23 percent) are relatively larger, and the share from transportation smaller (23 percent), than in the car-centric United States, according to the International Energy Agency. In any case, since power plants provide electricity to the other sectors, the direct burning of fossil fuels in those sectors can be reduced as they become more electrified.[2]

Thus, an overall decarbonization strategy must be built on five mutually reinforcing pillars. First, energy efficiency must shrink emissions from all sectors. Second, we must decarbonize electricity, the second leading source of emissions in the United States and the leading source globally. Third, we must use that clean electricity to replace fossil fuels for activities that can be electrified and use clean fuels for those that can't. Fourth, we must control potent greenhouse gases such as methane, nitrous oxide, and fluorinated gases from agriculture and other sources. Finally, a net-zero balance requires carbon sinks to offset any sources that remain.[3]

The five pillars reinforce each other in crucial ways. Efficiency limits demand for clean electricity and fuels, and thus the amount of production capacity and infrastructure that must be built to supply them. Clean electricity is needed to decarbonize not only existing uses but also newly electrified vehicles, heating, and industry. Demand for electricity will grow even larger if it is used to produce hydrogen and other "electrofuels" and power negative emissions technologies. Electrification enhances the efficiency of vehicles and heating. More subtly, electrifying vehicles and heating will transform the challenge of decarbonizing electricity itself, by shifting the timing and flexibility of power demand.

None of these pillars will be easy to construct. However, two of them—efficiency and the control of potent greenhouse gases—are straightforward. These are the win-win pillars of decarbonization, saving money and natural resources beyond their benefits to climate. If decarbonization were a weight loss plan, these two pillars would be as uncontroversial as eating less candy and cheesecake.

By contrast, two of the other pillars—clean electricity (Chapter 5) and electrification and clean fuels (Chapter 6)—will transform our entire energy diet. Like dieters puzzling over whether to go vegan or paleo, or adopt a Mediterranean or Atkins diet, the path forward is fraught with uncertainty. Sticking with the status quo is easy and might suffice with a few tweaks if the health of our climate had not already become so dire. But radical cuts require radical changes. Special care will be needed to ensure that our energy diet remains affordable and effective even as its ingredients are transformed.

To reach net zero, a healthier energy diet won't be enough. The dieter must also exercise enough to offset whatever calories remain.

That's where carbon sinks or "negative emissions technologies" come in (Chapter 7). Unfortunately, our analogy breaks down here, because most negative emissions technologies, unlike exercise, don't bring us side benefits beyond the offset itself. Unlike efficiency, clean electricity, and electrification, sinks don't cut our use of fossil fuels and all the environmental and health impacts that come with them. In fact, as we will see, some negative emissions technologies would consume enormous amounts of energy and other resources. Most negative emissions technologies remain costly or unproven at scale. Thus, the more that we can reduce emissions directly rather than offsetting them with sinks, the better off we are likely to be. Hopes for negative emissions technologies to emerge later should not quell the urgency of cutting emissions today.

Push and Pull Policies for Technology

Before turning to specific technology pillars, it is important to establish general principles for how policy influences their construction.

Technologies advance in three phases: *inventions* of new products or processes through research; *innovations* that commercialize those inventions, making them available in the marketplace through development and demonstrations; and the *diffusion* of innovations through broader adoption and deployment. Though boundaries between the phases are fuzzy, I'll use "R&D" (research and development) to denote the first two phases and "RD&D" to encompass all three phases, with the final "D" referring to diffusion via deployment.[4]

Without sufficient policy, the private sector tends to underinvest in RD&D for climate-friendly technologies for several reasons. First, those technologies yield environmental benefits that accrue to society at large rather than just producers and consumers. That creates a temptation to free-ride on the actions of others rather than invest in the public good. Second, technologies must cross one or more valleys of death, where they may languish for lack of funding to transform the inventions of basic research into profitable deployments at large scales. Huge investments may be needed to develop, prototype, and demonstrate technologies to cross those valleys. Valleys are especially deep for capital-intensive technologies such as nuclear power, geothermal drilling, and carbon-capture machines, which are far costlier to demonstrate and scale up than a software app. Even for technologies that survive, their developers and investors won't be alone in reaping the rewards. In fact, the spillover of technologies to other firms worldwide is crucial to broader diffusion of climate-friendly technologies, so long as it does not squelch the profit motivation to innovate.[5]

Overcoming these barriers to technology innovation and diffusion requires two forces. "Technology push" policies reduce the costs of invention and innovation, such as by funding or incentivizing R&D, and thus increase the supply of new technologies. A "demand pull" comes from policies that stimulate demand for those technologies, raising the rewards for innovation and diffusion. Government procurements and incentives stimulate demand directly, whereas emissions taxes, mandates, and standards do so indirectly by making fossil fuels competitors costlier or banned.[6]

Push and pull policies reinforce each other in essential ways. The technology push is especially important in stimulating the early stages of R&D, generating new technologies and driving down their initial costs. The demand pull can then accelerate diffusion of the technologies that emerge, further reducing costs and encouraging ongoing innovation.

Accelerated diffusion reduces costs thanks to the "learning by doing" that occurs as manufacturers learn to reduce costs and improve performance to remain competitive while scaling up. Learning by doing can be illustrated by plotting a curve of how the cost of a technology falls as its cumulative deployment grows. Such curves for LED lights, solar modules, and lithium-ion batteries show cost declines of roughly 20 percent for each doubling of deployment. For LEDs and solar, those "learning rates" have been enough to drive down costs by more than a factor of 100 across many doublings. Learning by doing breeds self-propelling cycles, as innovation and deployment drive down costs, stimulating further deployment. A technology push or a demand pull can each spur these cycles.[7]

On the technology push side, federal investments in R&D yield at least four times their cost in energy savings and even larger benefits to the environment. The federal government drives R&D through the work of national laboratories, funding for research at universities and other institutions, and incentives for corporate R&D. However, federal funding for energy RD&D totaled just 0.04 percent of GDP in 2019. That's just one tenth the share of GDP that the United States spent for the Manhattan Project to develop the atomic bomb or the Apollo Program to put a man on the Moon. Decarbonizing the economy to combat climate change is at least as important an undertaking. Reports issued by a National Academies panel, the American Energy Innovation Council, and Columbia University soon after the election of President Biden all recommended tripling federal funding for clean energy RD&D.[8]

The private sector should provide a stronger push too. The energy industry invests just 0.5 percent of revenues in R&D, compared with 14 percent in the pharmaceuticals industry. Venture capital firms poured $25 billion into cleantech start-ups from 2006 to 2011 but shut off the taps after losing more than half of their investments. As calls for net zero have intensified and drawn the attention of billionaires and industry, huge new venture capital funds have emerged, including the $2 billion Climate Pledge Fund created by Amazon and the $1 billion Breakthrough Energy Ventures led by Bill Gates. New entities such as Third Derivative, a joint venture of the Rocky Mountain Institute and New Energy Nexus that began funding fifty start-ups in 2020, aim to help fledgling enterprises cross the valley of death between demonstrating an initial proof-of-concept and scaling it up to profitability. Especially promising are public-private partnerships and cleantech incubators such as Cyclotron Road at Berkeley Lab, which trains recent doctoral graduates to translate their research into commercial endeavors. Since my visit to Cyclotron Road in 2018 while researching this book, participants I met have gone on to land tens of millions of dollars for their start-up ventures.[9]

One debate regarding technology push policies is the relative importance of pushing incremental improvements in existing technologies versus pursuing breakthrough innovations. The RMI co-founder Amory Lovins told me, "You can do everything you need to do to get to 1.5°C with 2010 technologies." By contrast, Bill Gates advocates a quest for "energy miracles," writing, "We need a massive amount of research into thousands of new ideas—even ones that might sound a little crazy—if we want to get to zero emissions by the end of this century."[10]

On the demand-pull side, federal, state, and local governments can directly boost demand for clean technologies through their procurements. Just days after his inauguration, President Biden vowed to replace the federal government's entire fleet of 650,000 vehicles with American-made electric models. The Climate Mayors Electric Vehicle Purchasing Collaborative has committed hundreds of cities along with counties, transit agencies, and universities to purchase electric vehicles. Incentives stimulate purchases by others. Cities and states are also among the largest purchasers of renewable electricity, and they often insist on highly efficient designs for their buildings.[11]

Incentives boost clean technology purchases by the private sector. Energy bills passed under Presidents George W. Bush, Barack Obama, and Donald Trump all featured billions of dollars in incentives to improve the efficiency of homes and businesses. Unfortunately, some research suggests that home weatherization incentivized by the 2009 stimulus bill yielded less energy savings than predicted by engineering models. Those models tended to overestimate both baseline energy use and the effectiveness of adding wall insulation. Poor workmanship may also have impaired results. Future weatherization policies should enhance worker training to ensure that cost-effective materials are properly installed. Special attention is needed to serve low-income households, which may lack money for upfront costs yet would benefit most from long-term savings.[12]

An emissions tax would create a demand pull by helping clean technologies outcompete their competitors. Most energy-economy models predict that even a hefty carbon tax would reduce energy use by less than 10 percent. However, many of those models treat technology as a given, regardless of policy. If the demand pull from emissions pricing, or any other policy, accelerates deployment, it can drive down costs along technology learning curves. Furthermore, if some revenue from emissions pricing is devoted to RD&D, it can provide a push toward technology improvements.[13]

Despite the merits of procurements, incentives, and taxes, the strongest demand pull comes from blunter tools—mandates and standards. When we really want to eliminate something as soon as possible, such as lead in gasoline or DDT in pesticides, we don't just tax it or incentivize its competitors—we ban it. If our aim is net zero by 2050, there simply isn't time to wait and see how people respond to taxes and incentives. Cars and appliances may be replaced only a couple of times by then, and buildings just once if at all. Every cycle of replacement of cars and appliances must bring a leap in performance. Old buildings must be retrofitted for greater efficiency, since most of today's homes and offices will still exist in 2050. Mandates and standards assure manufacturers that there will be market demand for zero-emission vehicles, carbon-free electricity, advanced biofuels, and cleanly produced steel, concrete, and other materials.

Energy Efficiency

Technology-push and demand-pull principles inform the pursuit of the first pillar of decarbonization, energy efficiency. Efficiency is the ultimate win-win pillar, yielding lower costs, reduced dependence on imports, and cleaner air and water, alongside its benefits to climate. Yet efficiency is too often overlooked. "People think of efficiency as tinkering around the margins, and not as the single most effective climate solution that it is," said Ben Evans of the U.S. Green Building Council.[14]

Apart from recessions and pandemics, energy use is perpetually driven up by population and economic growth, as more and wealthier people buy more products and travel greater distances, and driven down by energy efficiency, which lets us do more with less. For most of American history, population and economic growth outpaced efficiency, so energy use soared. Only since 2005 has efficiency strengthened and growth slowed to the point that energy use has plateaued. But in the absence of new policies, government forecasters project that energy use will rebound 10 percent by 2050, with economic growth outpacing anticipated efficiency gains.[15]

Most net-zero strategies, like those developed by researchers from Princeton University and the United Nations Sustainable Development

Solutions Network, depend on energy use falling by 20 to 30 percent to make decarbonization feasible. Reports by the American Council for an Energy Efficient Economy (ACEEE) and the Rocky Mountain Institute suggest efficiency could improve even further, cutting energy use in half by 2050. In all these visions, energy savings would come not just from what we traditionally think of as "energy efficiency," such as more efficient lightbulbs, appliances, insulation, and manufacturing practices, but also from another pillar of clean energy, electrification. Many electric devices use far less energy than the fuel-burning ones they replace. For example, electric motors are far more efficient than gasoline or diesel engines. Electric heat pumps use far less energy than gas or fuel oil furnaces. Only by pairing energy efficiency with electrification can energy use be cut most deeply.[16]

Just as electrification eases the path to efficiency, efficiency eases the path to the other pillars of decarbonization. Improving the efficiency of already electrified items, like appliances, and soon-to-be electrified ones, like cars and heaters, can reduce the amount of infrastructure that must be built for clean electricity. Improving the efficiency of items that can't be easily electrified, like airplanes, overseas ships, and certain industrial processes, minimizes the need for low-carbon fuels to power them and carbon sinks to offset their emissions. "Efficiency is very important to keep the costs of net zero down," said Steven Nadel, who led the ACEEE study.[17]

With so many opportunities for energy efficiency, one way to consider them is to create a "supply curve" of options in order of cost-effectiveness, showing how much energy could be saved at each price. Options that cost less than the energy they save are sometimes called "no-regret" opportunities, since they are profitable even apart from their environmental benefits. Energy efficiency supply curves were first plotted by scholars at Lawrence Berkeley National Laboratory in the early 1980s and popularized by a McKinsey and Company report in 2009. The McKinsey report catalogued no-regret opportunities for efficiency that could collectively cut U.S. energy use by 23 percent, while yielding twice as much savings as costs.[18]

Technology optimists argue that such supply curves understate opportunities for energy efficiency, for two main reasons. First, by charting only current conditions, efficiency supply curves neglect the improvements that arise when technologies are more widely deployed. Technology learning curves show that costs can fall dramatically as deployment grows.[19] Second, efficiency supply curves treat technologies in isolation. Amory Lovins argues that even more opportunities arise when individual technologies are combined into integrative designs. "How you combine technologies is as important as how good the technologies are," Lovins said, while showing me around RMI's innovation center in Basalt, Colorado. The center embodies integrative design, using ultra-efficient windows and high-performance insulation to eliminate the need for mechanical cooling and drastically cut heating needs. RMI cut the center's energy use so low that it is more than offset with a modest rooftop array of solar panels. Integrative design can boost the efficiency of vehicles too, using ultralight materials, better aerodynamics, and more efficient tires to reduce demand for energy from engines or batteries, enabling them to be lighter and cheaper as well.[20]

Countering technological optimism, some economists downplay opportunities for efficiency. How could there be money lying around waiting to be grabbed? If it were there, people would be grabbing it. Various obstacles obstruct their path. The long lifetimes of buildings, vehicles, and equipment limit the opportunities to replace them. Even when items are replaced, purchasers may not know about new technologies or may lack the upfront capital to invest in them. Researching new alternatives, finding contractors to install them, and other time-consuming steps can impose hidden opportunity costs that are neglected by technology optimists. Businesses often steeply discount future savings, adopting only options with payback times of a year or two even if savings would accrue for decades. Also, there are often mismatches between who pays the upfront costs and who enjoys the energy savings later. Builders and landlords may not invest in efficiency if it is homebuyers or tenants who will reap the savings. Even when energy-saving devices are purchased, they are not always used properly. For example, less than a third of households with programmable thermostats actually program their thermostats.[21]

Economists also note that two "rebound effects" may offset energy savings from efficiency. Direct rebounds occur when efficiency makes it cheaper to use an item, so we use it more. After switching to efficient LED bulbs, for example, I find myself leaving more of the lights on. Indirect rebounds occur when money saved from efficiency lets us spend more on something else. My energy savings at home may help me afford a flight to Europe, for example. Nonetheless, scholars estimate that direct and indirect rebounds collectively offset just 20 percent or so of energy savings from efficiency. "Rebound effects are small and are therefore no excuse for inaction," economist Kenneth Gillingham of Yale University has concluded.[22]

Given barriers such as lack of awareness, discounting of future savings, mismatched priorities, and rebound effects, it is clear that carbon taxes and incentives are insufficient to drive the unprecedented efficiency gains required for net zero. Mandates and standards are needed to ban inefficient and inferior products and establish cycles of ongoing improvement. "Standards really lock in savings and drive investments in technology," said Kateri Callahan, former president of the Alliance to Save Energy. The Energy Policy and Conservation Act of 1975 and subsequent legislation require the Department of Energy (DOE) to set and periodically update efficiency standards for more than sixty categories of products, such as air conditioners, dishwashers, televisions, and industrial equipment. The cycles of updates push manufacturers to meet or exceed each round of standards. DOE can then lock in the best performance as the baseline for the next round. That dynamic process has yielded more than $1 trillion of cumulative energy savings, ACEEE estimates.[23]

Updates to efficiency standards faced unprecedented hurdles during the Trump era. DOE withdrew its long-scheduled update to lightbulb standards and restricted future updates for other products, claiming that the previous administration had "misconstrued existing law." President Trump repeatedly bemoaned LED lighting for making his skin look orange, even though observers attribute his skin tone to facial bronzer. Libertarian think tanks launched a "Make Dishwashers Great Again" petition drive, falsely claiming that efficient models don't get dishes clean. Bowing to the requests of these groups and President Trump, DOE in 2020 weakened the water and energy standards for showerheads, washers, and dryers. President Trump even tried to nix EPA's Energy Star program, which informs consumers about energy-efficient products and yields far more energy savings than the program costs. Restoring bipartisan support for efficiency and warding off demagoguery and misinformation will be fundamental to getting efficiency back on track.[24]

Efficiency gains are needed most urgently in the transportation sector, the leading source of U.S. emissions. The United States has long lagged behind other countries in fuel economy standards for light-duty vehicles (cars, pickup trucks, sport utility vehicles, and minivans), which make up the majority of transportation emissions. Even the ambitious ramp-up scheduled under President Obama would have left U.S. standards trailing those in the European Union and major Asian countries. The rollback by President Trump left the United States further behind global norms. Many European countries have set plans to ban new gasoline and diesel cars by 2030, and manufacturers such as General Motors are planning to follow suit soon thereafter. Whether it results from an outright mandate or gradual adoption, a shift to zero-emissions vehicles will be crucial to achieving deeper progress than is possible with fossil fuels. Corporate commitments lock in progress even through political swings, making emissions reductions less vulnerable to each election.[25]

For freight trucks, the source of nearly a quarter of transportation emissions, fuel economy for the on-road fleet barely budged from the 1990s through the mid 2010s. However, regulations issued under President Obama and left unscathed by President Trump require new truck emission rates to step down about 40 percent from 2013 to 2027, a world-leading pace. Even steeper cuts will require switching trucks to run on electricity or hydrogen (as discussed in Chapter 6) and accelerating the replacement of old trucks.[26]

California is pursuing even stricter vehicle emission standards. Federal policy since 1968 has allowed California to set its own emissions standards, recognizing the challenges of improving air quality in the state's heavily populated, pollutant-trapping basins. Other states can adopt either California's rules or federal ones. During the Trump era,

California opposed rollbacks to light-duty vehicle standards and set bold rules for heavy trucks, requiring half of their sales to be zero-emissions by 2035, and all by 2045. As more and more states opt in to California's rules, their influence on trucking markets will multiply, creating a demand pull for clean trucks.[27]

For airplanes, the EPA waited until 2020 to propose its first greenhouse gas emissions standards and set them merely at levels already required by the International Civil Aviation Organization. The United States should instead establish cycles for updating standards for airplanes as well as ships, trains, and construction equipment, mimicking the process that has worked so well for appliances and industrial equipment.[28]

For new buildings, the International Energy Conservation Code is updated every three years by the International Code Council. After decades of stagnation, a series of ambitious updates doubled efficiency standards from 2006 to 2021. Unfortunately, most U.S. states have not kept pace with these model codes. By 2018, only a dozen had adopted residential building codes as stringent as the 2012 standards; eight states had no statewide codes at all. Commercial codes lag similarly behind. Since most buildings last for decades, weak codes lock in inefficiencies long after net zero should be sought. Building codes that took effect in California in 2020 require new homes to achieve zero net energy use through a mix of efficiency and solar power, yielding homeowners twice as much savings as costs. Zero net energy will be the standard for all new California commercial buildings starting in 2030, with a growing number of existing buildings retrofit to that standard as well. California's codes should become models for adoption by other states, leading to a transformation in how homes and offices are built.[29]

The United States faces unique challenges in pursuing energy efficiency where gasoline, natural gas, and electricity each cost far less than in Europe. Americans have therefore been slower than our peers to adopt energy-saving technologies such as LED lights and induction cooktops, which cost more upfront but yield savings over time. Urban sprawl and an *On the Road* culture of road trips fueled by cheap gasoline push Americans to drive nearly twice as far per capita as Europeans. Libertarian groups such as FreedomWorks have fueled opposition to lighting and appliance standards that once won bipartisan support. Energy prices have plunged in real terms since the last time major efficiency legislation was passed in 2005 and 2007. "What drove those bills was an energy crunch; it's very hard to get traction in times of abundant cheap energy," said Evans of the U.S. Green Building Council.[30]

Despite these challenges, the United States is uniquely positioned to improve efficiency domestically and catalyze progress abroad. Given that we use roughly 20 percent more energy per dollar of GDP than most other wealthy countries, we have an exceptional amount of fat ready to be cut from our energy diets. Cheap energy here has spurred innovators to design products whose virtues extend beyond their efficiency. It's little wonder that Tesla cars and Nest self-adjusting thermostats were developed in Silicon Valley and marketed more for their sleek designs and creature comforts than energy savings. "People will adopt a technology if it lets them do more, not just do things more cheaply," said Dane Christensen of the National Renewable Energy Laboratory.[31]

Energy-saving technologies designed to thrive in America's cheap-energy marketplace will be even more attractive in countries where energy is costlier. Thus, American ingenuity can be key to spreading efficient technologies beyond our borders, if given enough push from RD&D or pull from regulations. Technology transfer will be needed to make those innovations available beyond our borders. As the world's largest consumer market, efficiency standards set here can become a benchmark for manufacturers to meet worldwide, especially if the United States partners with other countries in climate clubs.

Methane

Methane, nitrous oxide, and fluorinated gases like chlorofluorocarbons (CFCs) and hydrofluorocarbons (HFCs) are emitted in tiny quantities compared to carbon dioxide, but their potency in trapping in Earth's heat makes them important to control. Methane is roughly 35 times as potent as carbon dioxide averaged over 100 years, with methane's impact concentrated in the first decade before atmospheric chemistry removes it. Nitrous oxide is nearly 300 times as potent as carbon dioxide, with impacts spread more evenly across the century. Fluorinated gases vary in their potency and lifetimes.[32]

EPA tallies methane, nitrous oxide, and fluorinated gases to be 10, 7, and 3 percent, respectively, of its U.S. greenhouse gas inventory for 2019. Globally, researchers from the Netherlands estimate the shares to be 19 percent methane, 6 percent nitrous oxide, and 3 percent fluorinated gases, with the rest carbon dioxide. However, both inventories understate methane's potency by about a third and may underestimate methane leaks from the oil and gas industry. Thus, methane's actual share of overall greenhouse gas emissions is likely larger. Methane and accompanying light hydrocarbons also contribute to ground-level ozone smog, so controlling them benefits air quality too.[33]

Fossil fuels are responsible for about half of methane emissions domestically and one-third globally in the 2019 inventories, mainly via leaks from oil and gas wells, natural gas systems, and coal mines; the actual fractions may be larger if leaks from oil and gas are underestimated. Using less fossil fuels would inevitably reduce these leaks. Leaks can also be tackled directly by completing oil and gas wells more carefully and replacing or retiring antiquated pipes. Poorly completed wells and antiquated pipes emit many times more methane than their better-performing rivals. The MethaneSAT satellite planned by the Environmental Defense Fund and SpaceX will enable methane leaks to be identified from space, complementing leak detection from infrared cameras on the ground.[34]

Since methane is the main component of natural gas, controlling leaks avoids wasting a valuable fuel. It also curbs the climate footprint of oil and gas, helping the industry market its gas as a clean alternative to coal. That framing will become increasingly tenuous as renewables supplant coal as the leading competitors to gas. Hence, most of the biggest oil and gas companies are voluntarily curbing leaks and lobbying for stronger regulations to ensure their competitors do too. The economic folly of Trump-era rollbacks of methane rules became clear in 2020, when the French government blocked Engie from importing $7 billion of U.S. liquefied natural gas because of its high emissions footprint. That could be a harbinger of how climate policies can influence the international competitiveness of fuels and products based on how responsibly they are produced and transported.[35]

Beyond oil and gas, methane leaks from coal mining can fall as the industry fades. Methane from landfills and wastewater treatment plants can be captured for use as energy. Better management of rice paddies can reduce methane emissions by up to 90 percent. For livestock, adding trace amounts of a red seaweed to their feed can reduce their burps and farts of methane. Livestock waste can be processed in oxygen-free tanks known as anaerobic digesters, which capture the methane for energy while producing biofertilizers.[36]

Nitrous Oxide

Nitrous oxide emissions come mainly from agricultural soils. Wasteful overuse of fertilizers adds more nitrogen to soils than crops can use. Some of that excess nitrogen enters the air as climate-warming nitrous oxide or as smog- and particle-forming nitrogen oxides and ammonia. The latter two compounds make agriculture the largest contributor to air pollution health effects in the United States, according to recent research. Excess nitrogen also washes away with rain, creating algal blooms in lakes and dead zones in coastal waters. The air, water, and climate impacts of excess nitrogen all can be minimized by applying fertilizers with what the U.S. Department of Agriculture calls the 4Rs—right source, right rate, right time, and right place. Doing so also improves crop yields while reducing fertilizer expenses. Further gains can be achieved with controlled-release fertilizers, which help crops use nitrogen more efficiently, and urease inhibitors, which slow nitrous oxide formation in soils.[37]

Despite these opportunities for affordable controls, agricultural emissions remain largely neglected. It's tough to regulate emissions that arise from millions of farms, few of which monitor their emissions. Without regulations or incentives, farmers have little reason to cut emissions. Captured methane has little value while natural gas is so cheap, except in locations such as California that have policies favoring renewable natural gas. Cheap fertilizers and old habits slow the adoption of better fertilizing practices. Thus, even as techniques and technologies improve, spurring their deployment will remain a daunting task.[38]

Our dietary choices and food waste affect emissions too by influencing what farmers and ranchers produce. A vegetarian diet cuts the emissions footprint of food in half by averting the burps and waste of livestock and the nitrogen fertilizers used to grow crops to feed them. Roughly a third of the nation's food supply is wasted, magnifying the emissions from any diet.[39]

Fluorinated Gases

Fluorinated gases such as chlorofluorocarbons (CFCs) that deplete stratospheric ozone were already phased out by the Montreal Protocol. But some of their replacements, like hydrofluorocarbons (HFCs), are extremely potent greenhouse gases too, even though they are safe for ozone. Refrigerant use is expected to soar as refrigerators and air conditioners become more available worldwide. As noted earlier, amendments to the Montreal Protocol, negotiated in Kigali in 2016, would phase down HFC emissions more than 80 percent by 2050 if fully implemented, averting up to 0.4°C of warming by 2100. Aggressive HFC policies will be crucial to spurring R&D and adoption of new technologies for cleaner refrigerants and the air conditioners and other devices that use them.[40]

Challenging as it may be, we know what we need to do to cut emissions of methane, nitrous oxide, and HFCs, just as we know what it will take to improve efficiency. How to build the remaining pillars of decarbonization is far less clear. Should we prioritize renewables, nuclear energy, or carbon-capturing fossil fuel plants as sources of clean electricity? How should we balance the variable output of wind and solar? Should cars and heating be converted to run on that clean electricity or use hydrogen fuel cells, advanced biofuels, or other cleanly produced fuels? How should hydrogen and other fuels be produced? What negative emissions technologies hold the most promise for offsetting the emissions that remain so that we can reach net zero? These questions and more form the basis for our explorations in the next three chapters.

five DECARBONIZING ELECTRICITY

Efficiency can take us only so far if vehicles, buildings, and industry continue to run on fossil fuels. But switching them to electricity makes sense only if electricity is clean, affordable, and reliable. That makes clean electricity the central pillar of decarbonizing energy. In fact, in most deep decarbonization scenarios, decarbonizing electricity would provide more emissions reductions than any other pillar.[1]

Power plant emissions in the United States have already fallen by a third since 2005, dropping below transportation as the leading sector. But what has carried us that first third cannot carry us the rest of the way to clean electricity. The leading sources of carbon-free electricity historically, hydropower and nuclear, have stagnated for decades and are ill-prepared to grow. Much of the progress since 2005 has come by replacing coal with natural gas, but gas without carbon capture can only get so clean. Wind and solar power have grown rapidly but remain dwarfed by fossil sources of power (Figure 6).[2]

Deeply decarbonizing electricity will require new paradigms for how power is generated, transmitted, and used. Contrary to some prior expectations, wind and solar now provide the most affordable sources of electricity. New options are emerging for balancing the variable output of wind and solar. Now that many states and utilities have committed to 100 percent clean electricity by 2050 or sooner and President Joe Biden has set that as a national target for 2035, new paradigms must be pursued right away to reshape the electric sector, which historically has evolved far more slowly.[3]

Hydropower

For nearly a century after commercial electricity generation began, clean electricity meant mainly one thing—hydropower. Hydropower began with a dam on Wisconsin's Fox River in 1882. Then came a building boom of dams during the Great Depression and another

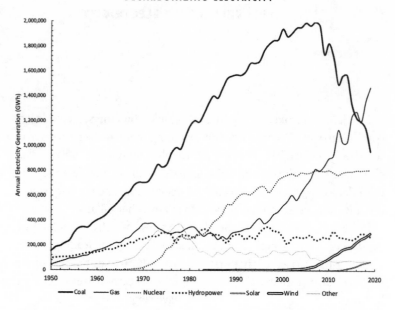

Figure 6. Electric power generation in the United States (Plotted by the author with data from Energy Information Administration, "Monthly Energy Review," 2020)

tripling of output after World War II. Calling traditional hydropower "clean" is a bit of a misnomer, since dams create reservoirs that flood the ecosystems behind them and can release large amounts of methane. By the 1970s, most prime locations for hydropower had been tapped, and environmentalists became increasingly effective at blocking construction of new dams. Dozens of dams are now being dismantled to restore river ecosystems, and no major hydropower projects are being built in the United States.[4]

The era of massive dam construction, flooding vast swaths of land with new reservoirs, is over. Any growth in hydropower must come with a lighter footprint. Some growth could come from upgrading equipment at existing hydropower plants or retrofitting unpowered dams to generate power. Together, that could boost nationwide hydropower output by around 10 percent, merely enough to keep pace with closures of some dams.[5]

What's less clear is how much hydropower could be added by building small-scale sites that do not require new dams. Turbines can be installed within rivers or partial diversions of their flow. However, environmental concerns could constrain opportunities. Studies by the Department of Energy in 2006 and 2014 identified thousands of potential small-scale sites that could collectively boost nationwide hydropower capacity by more than half. However, its subsequent *Hydropower Vision* study in 2016 found that only a handful of those sites lie outside "environmentally sensitive" regions. *Hydropower Vision* defined "environmentally sensitive" incredibly broadly, including any of seven wide-ranging traits. Thus, opportunities for small-scale projects may not be as scarce as *Hydropower Vision* implied but are limited nonetheless.[6]

Nuclear Power

As with hydropower, construction of nuclear plants surged before ultimately leveling off. After the devastation wrought by nuclear weapons in World War II, scientists moved quickly to harness nuclear energy for peaceful purposes. By 1954, the chairman of the U.S. Atomic Energy Commission was already predicting that nuclear power could become "too cheap to meter" within a generation. An ensuing building spree pushed nuclear beyond hydropower as the leading source of carbon-free electricity. In 1974, the Commission projected nuclear power capacity could reach up to 960 gigawatts by 1995, enough to supply most of the nation's power needs. But nuclear power never became cheap. By 1978, amid cost overruns and construction delays, the Commission had already scaled down its capacity projections by more than two-thirds.[7]

The following year, disaster nearly struck. A reactor at the Three Mile Island plant in Pennsylvania suffered a partial meltdown. The accident posed only minimal health risks, as radiation remained contained. But the jolt to public opinion was severe. Opposition to nuclear power mushroomed. No new nuclear plants were completed in the United States in the four decades that followed. Closures of some reactors have pulled U.S. nuclear capacity back below 100 gigawatts, just one-tenth of the Commission's wishful projections from the 1970s. Similarly, global nuclear capacity has grown at less than one-tenth of the pace that the International Atomic Energy Agency predicted in 1974. Only about fifty nuclear reactors, mostly in Asia, were under construction in 2021, barely enough to keep pace with retirements.[8]

With construction stalled, utilities have managed to keep aging nuclear plants running longer and more productively than originally planned. Nuclear plants were originally designed for an expected lifetime of twenty-five to forty years. All remaining U.S. plants have aged well beyond that range, earning license renewals from the Nuclear Regulatory Commission twenty years at a time. Nuclear plants now need far less downtime, operating at over 90 percent of their capacity compared with just 50 percent in the early 1970s. Those capacities have also grown as utilities have upgraded their reactors.[9]

All of these developments helped nuclear reach and maintain a plateau of nearly 20 percent of the U.S. electricity supply through the 2010s. However, that share is set to fall, as eight of the nation's ninety-four remaining nuclear reactors are scheduled to close by 2025. Others are grappling with falling prices for power, rising maintenance costs, and public opposition. Bailouts of reactors by the states of Illinois and Ohio were tainted by bribery schemes.[10]

The Energy Policy Act of 2005 sought to jump-start construction of new nuclear plants, motivated by natural gas prices that were spiking at the time. Generous loan guarantees and other incentives spurred developers to propose thirty-one new reactors. But interest faded when natural gas prices collapsed amid the fracking boom. By 2016, only two pairs of reactors were under construction. South Carolina halted construction of the pair in its state after projected costs ballooned to more than twice the original estimates; the utility's vice president subsequently pleaded guilty to fraud and lying about construction progress. The other pair, in Georgia, remains years behind schedule and more than $10 billion over budget. Together, the projects bankrupted the lead contractor, Westinghouse, and left the states' utilities and ratepayers saddled with enormous expenses. Rather than being too cheap to meter, nuclear has become too expensive to build and operate. With so few reactors being built, the nuclear industry has not enjoyed the learning by doing that has driven down the costs of its competitors. Only if new designs for large-scale reactors prove successful and affordable overseas are we likely to see more large-scale projects built in the United States.[11]

While the debacles in South Carolina and Georgia stifle interest in building large-scale nuclear plants, a handful of companies are seeking to deploy smaller reactors. They have designed small modular reactors (SMRs) or even smaller microreactors that could be mass-produced in factories, shipped by truck or rail, and installed in multiples on-site. "If nuclear is to become cheap, it needs to go to factory fabrication," nuclear expert Jessica Lovering has said.[12]

The first major deployment of SMRs is likely to come from NuScale, which plans to install an array of 77-megawatt units at a site in eastern Idaho by 2029. TerraPower, founded by Bill Gates, hopes to deploy its first 345-megawatt sodium-cooled SMR in Wyoming, pairing it with molten salt heat storage to provide flexible power output to balance wind and solar. X-energy is designing SMRs to produce high-temperature steam for industry when electricity demand is low. Another nuclear start-up, Oklo, is designing a microreactor with a capacity of just 1.5 megawatts, less than a single wind turbine, to replace diesel generators in remote locations.[13]

Costs, safety, and waste remain key challenges, even with these modular designs. Costs per kilowatt-hour will be even higher than traditional nuclear plants at first, since SMRs and microreactors will not operate as efficiently or spread their costs over as much output. Reducing costs over time would require scaling up manufacturing. Scale-up won't come quickly, slowed by environmental and safety reviews and the novelty of deployment. As various companies pursue starkly different designs, no single design is poised to scale up soon. As of mid 2021, only NuScale had announced both a site and target date for its initial U.S. pilot project. Thus, widespread U.S. deployment of any new nuclear technologies before the mid 2030s seems unlikely, limiting the role that they can play in meeting clean electricity targets.[14]

As to safety, NuScale claims that advanced safety features, such as fewer moving parts and the ability to shut down automatically, would make its SMRs "as safe and simple as you can get." Thus, the company has asked the Nuclear Regulatory Commission to relax some of its safety requirements, such as the size of the emergency evacuation zone. That would expand where SMRs could be installed, but the proposal alarms critics. "To say that you know so well how a new reactor will work that you don't need an emergency evacuation zone, that's just dangerous and irresponsible," physicist Edwin Lyman of the Union of Concerned Scientists told *Science*. Meanwhile, Oklo is seeking uranium enriched up to nearly 20 percent, several times what a typical reactor uses, so that its microreactors could operate for years without refueling. No such fuel is available on the market today, and with good reason. According to the Government Accountability Office, such fuel would be "a more attractive target for theft or diversion into a weapons program, because less work is needed to make it into weapons-grade uranium." Even if the risks of proliferation of microreactor fuel are small, incurring them for the sake of such tiny sources of power is unwise.[15]

If nuclear power ever does scale up, whether in modular or traditional forms, so would nuclear waste. Industry advocates tend to be dismissive of concerns about waste. From a technical perspective, that's reasonable. "Technically it's just a matter of finding a deep enough, geologically stable hole," Alex Gilbert of the Nuclear Innovation Alliance told me.[16]

But politicians have yet to muster the will to build any of those holes, not just in the United States but globally. Instead, nuclear waste remains the world's most solvable problem that's never been solved. Only Finland has begun construction of a long-term waste repository. In 1987 Congress chose Yucca Mountain in Nevada as the site for the first American repository. But construction never began due to opposition from the state's politicians. No alternative site has been found. That has forced nuclear plants to store their own spent fuel, first in cooling ponds and then in dry casks, for far longer and in larger quantities than ever intended. They already store enough waste to need two Yucca Mountain–sized repositories, without a single repository on the horizon. "Most utilities are unlikely to build new nuclear until there is a long-range plan that does not include leaving them to manage the waste for centuries at their sites," nuclear expert Michael Ford told me. Modular reactors would operate less efficiently than larger reactors, producing more waste per unit of output. Transporting that waste to repositories could be challenging, if reactors are dispersed across many sites. "There are many states who will challenge transport through their state, making this a significant risk for delay," Ford said.[17]

We should maintain the clean power output of existing nuclear plants, upholding rigorous regulations for safety and maintenance and building long-overdue repositories for permanent storage of their wastes. Unfortunately, those aging plants will not last forever, and traditional designs have become too costly to build. SMRs hold promise for their modular designs and flexible output, but costs and performance remain uncertain and long-term waste disposal remains unresolved. That makes new nuclear technologies worthy of funding for research, development, and carefully monitored initial deployments, but not something that can be relied upon to decarbonize electricity anytime soon.

Carbon Capture from Coal and Gas

The same Energy Policy Act that tried to revive nuclear power also funded efforts to capture carbon emissions from coal plants. At the time, capturing emissions seemed more viable than replacing coal, which was then the leading and cheapest source of power. Most scenarios that had been considered by the Intergovernmental Panel on Climate Change expected wind and solar to remain far costlier than coal throughout most of the century and therefore to remain mere slivers of the power supply. Carbon capture was expected to play a far larger role.[18]

Trouble is, carbon capture from coal and gas never became affordable. At existing plants, capturing carbon dioxide requires absorbing it into a liquid and then releasing it as a concentrated gas to compress and send to storage. That requires energy to cool the exhaust, then to heat the liquid, and then to compress and transport the carbon dioxide. Thus, more fuel may be needed to maintain power output, pushing up needs for mining, transport, and waste disposal, and all the harms that come with them.[19]

Only two coal plants, including the Petra Nova project at the W.A. Parish power plant southwest of Houston, have installed carbon capture at commercial scales at existing facilities. Petra Nova touted itself as being built on time and on budget as it hosted delegations of journalists, congresspersons, and even my own students. But Petra Nova's claim of "transforming the industry" was unfounded. No companies, including the developers of Petra Nova, plan to replicate its approach. Petra Nova's viability depended on special circumstances, including $190 million in grant funding from DOE and a 50/50 stake in oil recovered by pumping its carbon dioxide into a depleted oil field. Much of the emissions reduction was offset by the natural gas plant that was built to provide power and steam for the carbon capture. Counting the recovered oil would wipe out environmental gains. As oil prices plunged during the COVID pandemic, facility operators quietly mothballed their carbon capture equipment, resuming electricity production with emissions uncontrolled. The mothballing came just after Petra Nova passed its three-year minimal obligation under the DOE grant, which had been issued with hopes for at least twenty years of operation. Even while Petra Nova was operating, uncontrolled sulfur dioxide emissions from the other coal boilers at Parish made it the deadliest power plant in Texas, research by my group has shown.[20]

Carbon capture from existing coal plants could be improved by using novel nanomaterials that soak up carbon dioxide more efficiently than amines. However, separating dilute carbon dioxide from exhaust and then compressing it inevitably requires lots of energy and expense. Also, clean electricity targets won't leave room for power plants that capture just a fraction of their emissions. Since old coal plants are hazardous environmentally and barely viable financially, replacing them with cleaner sources is the better route, at least in countries like the United States where coal plants are decades old and alternatives are readily available. If carbon capture retrofits make sense anywhere, it will likely be in parts of Asia where coal plants are far newer and gas and renewables are less abundantly available.[21]

Rather than retrofit existing coal plants to capture some of their carbon dioxide, other DOE-backed projects have sought to design new coal plants to capture nearly all of their emissions. One such approach, integrated gasification combined cycle (IGCC), would gasify coal to separate carbon and impurities before burning the remaining gas. That was the approach attempted by the original FutureGen project, launched by President George W. Bush in 2003. After years of delays and cost overruns, DOE abandoned construction in 2010. However, DOE continued to fund efforts by Mississippi Power to adopt that approach at Plant Kemper, which began construction that same year. The utility pulled the plug on those efforts in 2017, after falling three years behind schedule and $4 billion over budget. It instead converted Plant Kemper into a natural gas power plant without carbon capture.[22]

An alternative design, known as oxyfuel combustion, would burn coal in oxygen and recycled flue gas, concentrating carbon dioxide to make it easier to capture. However, it takes a lot of energy to separate oxygen from nitrogen in air. The first pilot-scale oxyfuel plant opened in Germany in 2008. Extending oxyfuel combustion with carbon capture to commercial scale was the aim of FutureGen 2.0, launched by DOE soon after it abandoned its original IGCC-based FutureGen. But DOE canceled FutureGen 2.0 in 2015 after years of delays and cost overruns. President Obama's pledge to open five to ten commercial-scale carbon capture plants by the end of his presidency went unfulfilled.[23]

NET Power is pursuing oxyfuel combustion with natural gas rather than coal, pioneering a novel approach known as the Allam cycle. After natural gas is combusted in pure oxygen, the resulting carbon dioxide rather than steam serves as the working fluid, turning the turbine to generate electricity and then being captured. When I visited their pilot plant near Houston in 2018, NET Power officials told me that they expected the technology eventually could match the efficiency and cost of the world's top natural gas power plants. That remains to be seen, since by early 2021 NET Power had announced only tentative plans for its first two commercial-scale plants.[24]

In sum, neither of the two historic leaders of clean electricity, hydropower and nuclear, nor a much-anticipated future one, carbon capture, is likely to grow substantially this decade. Each requires long lead times and enormous upfront costs, along with next-generation technologies that have not been proven at scale. Nuclear power inevitably requires mining, processing, transporting, and disposing of radioactive materials and warding off proliferation risks and public opposition. Fossil power inevitably requires mining or drilling, processing, and transporting coal or gas and mitigating the water and air pollution and other wastes that come from burning them, even if the carbon is captured. Yet targets for 100 percent clean electricity are approaching fast. That has left a huge void, which only wind and solar are ready to fill affordably, with help from complementary resources.

Wind and Solar

Long dismissed as costly niche options, land-based wind and utility-scale solar have suddenly become the most affordable sources of electricity (Figure 7). From 2009 to 2020, the cost of land-based wind power fell by 71 percent and solar by 90 percent. That has brought their costs to less than one-quarter of the cost of new nuclear or carbon-capturing coal plants and made solar the cheapest source of electricity in history. Costs for offshore wind are plunging too.[25]

For land-based wind, after a century of incremental progress, affordability and performance raced ahead over the past two decades. Average blade lengths more than doubled, letting turbines sweep through four times as much area. The hubs around which those blades spin climbed 30 meters (100 feet) higher into the air, reaching the height of the Statue of Liberty's torch, where winds blow more strongly and steadily. Advanced materials enable the blades to withstand the stronger forces that these winds and sizes entail. Economies of scale held down costs as deployments grew. Utilities tweaked the operation of turbines to produce steadier output and protect birds and bats. Thus, even as their capacity has grown, turbines now output an average of 42 percent of their capacity throughout the year compared to just 24 percent two decades ago. Along with better day-ahead wind forecasts, that has made wind output less intermittent and more predictable. Although conventional turbine blades can get iced, as in the Texas blackouts of 2021, winterized turbines can perform well even in the frigid conditions of Denmark or Antarctica.[26]

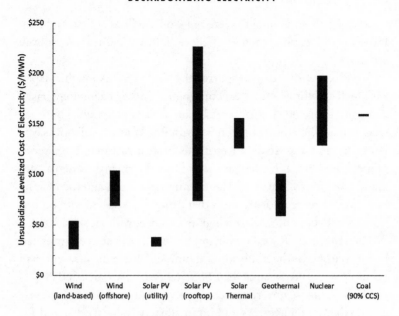

Figure 7. Unsubsidized levelized cost of electricity from carbon-free sources and from coal with 90 percent carbon capture. Rooftop solar reflects the range of residential and commercial. Solar thermal includes storage. Coal includes the cost of capturing and compressing but not storing the carbon. (Plotted by the author with data from Lazard, "Levelized Cost of Energy and Levelized Cost of Storage—2020," October 19, 2020)

Offshore wind turbines reach even higher and wider than land-based ones. Though twice as expensive as land-based wind, their costs are falling fast. That's making offshore wind increasingly attractive in coastal regions of Europe and the northeastern United States, where population density is high, land is scarce, and winds over the ocean far outpace those over land. Since the power of wind increases with the cube of its speed, faster wind speeds can dramatically boost output.[27]

Costs for solar have plunged even faster than for wind. Soon after their invention at Bell Telephone Laboratories in the 1950s, solar panels cost more than $100 in today's dollars per kilowatt-hour of electricity they produced. That was no match for other options, whose costs are measured in pennies or dimes. Thus, solar was viable only where all other options were not—on satellites at first, and then for off-grid locations such as offshore lighting and remote railway crossings.[28]

The oil crises of the 1970s spurred the United States and then Japan to launch ambitious solar R&D programs. Those "technology push" programs drove down solar costs by an order of magnitude, but they were still nowhere near competitive with fossil fuels. As oil prices collapsed in the 1980s, so did investments in solar research. But a series of incentives in California, Japan, and then Germany created a "demand pull" for solar panels. Manufacturers in Japan and then China built ever larger factories to meet that demand. Costs fell with economies of scale but remained too high to compete without subsidies.[29]

Then the Great Recession hit, just as some European countries had begun reining in their subsidies. Chinese manufacturers, who had been exporting nearly all of their output, saw overseas demand evaporate. The Chinese government stepped in with lines of credit to rescue reeling manufacturers and jump-started domestic demand for their products with hefty subsidies. Chinese installations soon soared, far surpassing the rest of the world combined. The scale-up in manufacturing drove down costs by another order of magnitude, letting solar catch up with wind as the cheapest sources of electricity.[30]

The extraordinary decline in solar photovoltaic costs came without any singular technological breakthrough. Most solar cells continue to use crystallized silicon in designs remarkably similar to those invented in the 1950s. That's largely a quirk of history. "If you asked a physicist to design a solar cell from scratch, they would never pick silicon," solar researcher David Moore of the National Renewable Energy Laboratory (NREL) told me. Silicon is not the strongest absorber of sunlight and is sensitive to defects, he said. "We use silicon because it's what Bell Labs came up with in the 1950s." Other photovoltaic materials and designs have come and gone without winning much market share. Relentless incremental progress in efficiently manufacturing traditional silicon cells, pulled by market demand, has won the day. As Ian Maxwell put it, "Never bet against silicon." Indeed, of the scant 1 percent of revenue that the solar industry spends on R&D, most aims merely to tweak existing silicon technologies. So far, bets on silicon have paid off. "Cheap solar is real, it is stunning, and I think it is the most important good news about climate of the 30 years that I've worked on the climate energy problem," David Keith of Harvard University has said.[31]

Unfortunately, even as wind and solar have become cheap, they grew to just 8 percent and 3 percent of the U.S. power supply, respectively, by 2020. Most net-zero strategies envision explosive growth in solar and wind, enabling them to supply roughly half of the nation's electricity by 2030 and 90 percent by 2050, even as electricity demand may double to power more vehicles, heat pumps, industrial processes, and green hydrogen production. As cheap as wind and solar have become relative to other clean electricity, they must become even cheaper to make electrification and green hydrogen more attractive. "Silicon solar panels may not lead us into a green economy where power is doing much more than it used to," said Varun Sivaram, the author of *Taming the Sun*.[32]

If solar cells could instead be printed onto a flexible material, spooled onto rolls, and then unrolled for installation, they could be deployed at unimaginable scales. That is the vision for an emerging technology known as perovskites. "If we could print solar cells like newspapers, we could speed up production by a factor of 100," said Moore as he showed me around his solar perovskite lab at NREL in Colorado. Moore envisions flexible perovskites being deployed on all sorts of surfaces where rigid panels cannot go, such as clothing, cars, and cell phones. However, current perovskite materials contain toxic lead and can degrade within months. Other emerging technologies, such as solar cells made from organic compounds or quantum dots, are also promising but beset by challenges. Some analysts question the need for new solar technologies. "Crystalline silicon technology is good enough, and it's hard to beat," Jenny Chase, Bloomberg NEF's head of solar analysis, told *Nature* in 2019. Perovskites, she said, "are not something we need to wait for." Whether new technologies emerge or crystalline silicon continues to dominate, what's clear is that solar photovoltaic energy is already cheap and getting cheaper.[33]

While sprawling solar farms now dominate the market, residential and commercial rooftops deserve renewed attention. In the United

States, costs for rooftop solar have not fallen nearly as fast as for solar farms, since marketing and installation remain expensive even though panel prices have plunged. That has left rooftop solar several times as costly as solar farms here. However, rooftop solar avoids the land use and ecosystem disruption of solar farms. It also provides power directly to the consumer, avoiding transmission and distribution that can be costlier than generation itself. So rooftop solar only has to narrow the gap to be valuable. If it does, rooftop solar paired with local storage could reduce overall costs for clean electricity, by shrinking the need for new solar farms and transmission.[34]

Policy can make rooftop solar more affordable. Rooftop solar in Australia, Mexico, and some Asian countries costs just a third as much as in the United States, thanks in part to more streamlined processes for approving and inspecting projects and connecting them to the grid. Low-interest financing and incentives from utilities or government can help middle-income families enjoy savings from rooftop solar. Deeper savings could be achieved by requiring solar on new homes, with exemptions where needed, as California began to do in 2020. That allows rooftop solar to be incorporated into a home's design, financed within a low-interest mortgage, sold without door-to-door marketing, built with economies of scale, and operated for as many years as possible.[35]

Despite the success of silicon solar panels and the emergence of promising alternative photovoltaic materials, solar thermal technologies have struggled. Unlike photovoltaics, which convert sunlight directly to electricity, solar thermal facilities generate heat by concentrating sunlight with sun-tracking mirrors. Those "heliostats" need direct beams of sunlight and thus are exceptionally dependent on clear sky conditions. Some facilities concentrate the sunlight in troughs, and others reflect it toward towers. The heat is used to produce steam to turn a turbine and generate electricity, either immediately or by storing heat in molten salts to produce steam and power later.

As recently as a decade ago, solar thermal was expected to become more affordable than photovoltaics, at least in exceptionally sunny regions. However, while solar photovoltaics were becoming affordable along learning curves thanks to factory production of modular designs, the one-at-a-time construction projects needed for solar thermal suffered the same sorts of delays and cost overruns that plagued nuclear and carbon capture projects. Each solar thermal site had to be designed around unique local conditions and faced unexpected operational challenges. At Ivanpah Solar in California, where over 170,000 heliostats reflect sunlight onto three towers, jet contrails obstructed the sun and misaligned mirrors set fire to one of the towers. At Crescent Dunes in Nevada, where more than 10,000 heliostats reflect sunlight onto a single tower, the utility halted operations after costs soared and leaks of molten salt and other mishaps crimped production.[36]

There are reasons to hope that solar thermal can overcome its recent woes and improve its performance and costs. Designs with molten salts can continue producing power after sunset, satisfying evening demand while photovoltaics cannot. About two dozen solar thermal projects are under development worldwide, so learning by doing may emerge if replicable designs are adopted. Still, like nuclear, solar thermal faces a race against time to prove its worth as a complement to wind and photovoltaics.[37]

Geothermal

Though often overlooked, geothermal energy could become a leading new complement to wind and solar, if emerging technologies pan out as hoped. That's a big if. But the prospect is less far-fetched than some people realize. In fact, it's likely there will be more additions of geothermal in the United States over the next two decades than of hydropower, nuclear, and coal combined. That may sound implausible, since for now geothermal supplies just 0.4 percent of U.S. electricity, mostly from wells built decades ago in California and Nevada. Those wells sit atop reservoirs that are exceptionally hot at shallow depths—so shallow that steam emanates naturally from the surface. Sites with such shallow heat are rare. Most geothermal resources lie several kilometers below the ground—too deep to justify drilling into them vertically. However, veering the drilling horizontally, via techniques pioneered for oil and gas, would tap into far more heat at once, boosting output and affordability. If those techniques succeed in pilot projects, they could be replicated at dozens of sites to drive down costs.[38]

Tim Latimer founded Fervo Energy to do just that. "We're going to be the first in the world to do horizontal drilling for geothermal, which we believe can be transformative," he told me. Fiber-optic sensing will help optimize flow rates once a geothermal reservoir is reached. Latimer used a two-year stint at Berkeley National Laboratory's cleantech incubator, Cyclotron Road, to develop his concepts and then landed funding from Breakthrough Energy Ventures to start his business. Next, he moved his company to Texas, realizing that the expertise he needed was not just in Silicon Valley but in the roughneck world of oil and gas drilling. The aim is not to be as cheap as wind or solar farms but to outcompete other potential complements to their variable output. "Even a resource that might be more expensive on a raw energy basis like geothermal plays a critical role in terms of delivering power when it's needed, so it becomes part of a least-cost solution to achieving deep decarbonization," Latimer said. Other start-ups such as Eavor Technologies of Canada and GreenFire Energy in California are pioneering their own designs for tapping geothermal energy with horizontal drilling.[39]

DOE's GeoVision study in 2019 envisioned that pairing horizontal drilling with other advanced technologies could enable geothermal power to generate 16 percent of U.S. electricity by 2050 at competitive costs. In 2020, based on pre-pandemic conditions, geothermal was estimated to cost twice as much as wind and solar and half as much as new nuclear plants. Plunging oil prices amid the COVID pandemic left hundreds of drilling rigs and thousands of workers idle. Together with low interest rates, that could slash the cost of geothermal drilling. Merely narrowing the cost gap with wind and photovoltaics would make geothermal a valuable complement, since it can provide continuous or flexible output resilient to any weather.[40]

Still, geothermal developers face numerous obstacles. Geothermal has received smaller and less consistent tax credits than wind and solar. Only one site, DOE's Field Observatory for Research in Geothermal Energy in Utah, provides a testbed for experimenting with unconventional techniques on a geothermal reservoir. "Basically, the entire future of geothermal in the United States right now comes down to whether that one site can perform," Latimer said. "Dozens of promising geothermal technologies that are ready for field testing are stuck." And while those technologies remain untested, utilities are unwilling to incorporate geothermal into their long-range planning. "To make geothermal energy cheap, we need 5 or 10 FORGE projects to sufficiently build on technological developments and to fail, innovate, succeed, and demonstrate—bringing down costs along the way," Erik Olson of the Breakthrough Institute has written. To allow successes to proliferate, Congress should reform the National Environmental Policy Act to expedite environmental review and permitting for geothermal projects, which currently face more onerous barriers than oil and gas drilling. With adequate investments in RD&D and an easing of regulatory barriers, geothermal could quickly become a leading complement to variable wind and solar. "Geothermal fills a critical gap to complete the energy transition," Princeton University energy modeler Jesse Jenkins told *Quartz* magazine.[41]

Oceans and Other New Sources of Electricity

Less attention has been paid to ocean-based sources of electricity, but they could play a niche role near densely populated coastal regions. Various studies have estimated that energy from waves, tides, and ocean currents could together supply more than 15 percent of U.S. electricity. However, technologies to harness that energy remain untested at commercial scales. Waves and tides are strongest along the Pacific and northeast Atlantic coasts, Hawaii, and Alaska, while ocean currents are strongest offshore from the southeast Atlantic coast. However, ocean-based electricity is for now far costlier than land-based wind and solar where those are available.[42]

Various other options for power generation have been proposed over the years, such as nuclear fusion or beaming down microwave energy from solar-collecting satellites. However, any new technology will need to compete with increasingly affordable wind, solar, and geothermal power and existing hydropower and nuclear power, and it must scale up in time for clean electricity targets.

A New Paradigm for Electricity

The shifting fortunes of each carbon-free option suggest that electricity should be decarbonized by a far different paradigm from what once seemed most feasible.

The plunging costs of wind and solar photovoltaics and the cost overruns of nuclear and carbon capture have caught experts and policy makers by surprise. In 2005, DOE's Energy Information Administration (EIA) forecast that wind and solar costs would barely budge through 2025, with solar remaining far costlier than nuclear or carbon capture. Thus, when Congress passed the Energy Policy Act later that year, there was little surprise that it targeted incentives toward nuclear and carbon capture from coal. A few years later, Gregory Nemet, now a professor studying low-carbon innovation at the University of Wisconsin–Madison, began interviewing dozens of solar experts about their expectations for how low solar costs could fall by 2030. By 2019, actual prices had already fallen below the experts' most optimistic forecast, Nemet told me.[43]

By 2013, when Jim Williams led the U.S. team of the Deep Decarbonization Pathways Project developing scenarios for clean energy, EIA had extended its flat cost projections for wind and solar photovoltaics to 2040, and it did not anticipate the further cost overruns that continue to plague nuclear and carbon capture. Thus, the Pathways team posited four pathways to clean electricity that Williams tells me seemed similarly plausible at the time: a High Renewables case, a High Nuclear case, a High Fossil Fuels with Carbon Capture case, and a Mixed case blending the other three. Skewed by EIA's pessimism on wind and solar, the 2013 study projected that the High Renewables case would be the costliest of the four.[44]

Since the original Pathways study was conducted, wind and solar costs have plummeted to one-quarter the cost of new nuclear or fossil fuels with carbon capture, and they are expected to fall further. In updated modeling with the latest cost projections in 2020, Williams and colleagues from the Zero Carbon Consortium found that an electricity generation mix with 90 percent wind and solar would minimize costs. Similarly, Princeton University researchers relied on wind and solar to provide 85 to 98 percent of the power supply in most of their scenarios for reaching net zero by 2050. "It's hard to justify high nuclear scenarios now," Williams told me. Also, with wind and solar making clean electricity cheaper than previously imagined, the newer studies envision more electrification of vehicles and heating, shrinking the demand for liquid and gaseous fuels.[45]

Power grids must maintain a perpetual balance between supply and demand. Making variable wind and solar the dominant sources of electricity will require a new paradigm for maintaining that balance. "The advent of renewable energy being very, very cheap means that our source of electricity will be variable by nature and far from where we live," said Christopher Clack, an electricity systems modeler who founded Vibrant Clean Energy.[46]

Electricity has traditionally come from a limited number of power plants, rather than thousands of wind turbines and solar farms. Traditional plants produce electricity either continuously, like nuclear and coal plants operating near their maximal capacity around the clock, or flexibly, like dams and gas turbines that can easily modulate their output. The continuous sources are often conceptualized as providing a firm "baseload" of power, with the other sources flexing their output to match fluctuating demand.

Sprinkling a bit of wind and solar into the mix is no problem. Wind and solar have grown to provide 11 percent of the power supply nationally, and far more in some regions like the wind-swept Great Plains, without interrupting the reliability of power. However, for wind and solar to become the leading sources of electricity, as their low costs and zero emissions merit, we will require a new paradigm for power supply. Once built, wind and solar are virtually free to operate, both in dollars and emissions. Thus, we will want to use them as much as possible. They, not traditional baseload plants, will serve as the foundation of the power supply. But it will be a bumpy foundation, shaped by winds and sun. Other sources must flex their output atop that bumpy base to balance variable demand.[47]

Most clean complements to wind and solar—hydropower, nuclear power, geothermal, and perhaps someday ocean-based sources or nuclear fusion—operate nearly continuously to dilute their upfront costs over as much output as possible. Flexing to balance wind and solar may reduce that output, although developers such as TerraPower for nuclear and Fervo for geothermal are designing their projects to provide that flexibility. Other complements, like biomass or renewable gas plants, may cost more to operate but flex their output more readily. Balance is a matter of not just supply but also demand. Demand tends to follow predictable patterns—rising when people come home to cook, dipping when they sleep, and peaking on the hottest afternoons or coldest mornings when air conditioners or heaters are running full blast. Unfortunately, winds tend to be slow on hot summer afternoons, and sunlight is weak on cold winter mornings.

Time-varying prices can incentivize users to shift demand away from times of scarcity toward times with more abundant wind and solar power. Some industries are adept at shifting demand to avoid peak prices, but residential consumers tend to be less responsive on their own. Nonetheless, smart thermostats, water heaters, and other devices that automatically adjust to grid conditions can reduce residential peak demand by at least 30 percent. Incentives and override options may encourage consumers to adopt such devices.[48]

Alongside complementary sources and more flexible demand, two other factors can keep power reliable and affordable on a grid dominated by wind and solar: transmission and storage.

Transmission

Complements to wind and solar, like wind and solar themselves, are geographically constrained. River flow, geothermal resources, ocean conditions, proximity to carbon storage sites, and public acceptance of nuclear all vary by region. Wind and solar need affordable land near power lines in windy or sunny areas. Output from all these sources varies with time, yet power supply and demand must be balanced continuously. Since electricity travels at the speed of light, it can be transmitted from where it's generated best to where it's needed most at any time of day. The broader the geography over which we can pool these resources, the more robust our supply can be. That's where transmission comes in.

As the saying goes, it's always five o'clock somewhere. As electricity demand reaches its late-afternoon peak on the east coast, the western United States is basking in midday sunshine. As west coast demand reaches its own peak, evening winds pick up across the prairies. Coastal winds peak with summer sea breezes. Nuclear, hydropower, and geothermal output can be fixed or flexible depending on how facilities operate.

Unfortunately, U.S. transmission grids are ill-equipped to link these resources. The American Society of Civil Engineers rated the grid a C– on its 2021 Infrastructure Report Card. Most transmission lines were built in the 1950s and 1960s, when centralized power sources dominated supply. Many of the nation's windiest and sunniest regions lack the high-voltage lines needed to transmit power efficiently. High voltages are key because they allow far more power to be transmitted with less wasted along the way.[49]

Transmission in the continental United States is balkanized into three main grids and further subdivided into regional systems managed by different operators with different rules and priorities. The western and eastern grids are split by a seam that runs along the Colorado-Kansas border up through eastern Montana. Texas has a grid unto itself, stranding its residents during the February 2021 blackouts. These grids and their regional systems have too few high-voltage transmission lines traversing windy and sunny areas.

Building high-voltage lines through windy and sunny areas can avert transmission bottlenecks that sometimes force grid operators to "curtail" the output from wind and solar farms. "We can build a lot less capacity if we invest in transmission, because we're not going to waste as much electricity," said Clack. He estimates that a "supergrid" of transmission lines would shave about $1 trillion off the cost of decarbonizing U.S. electricity, since its costs would be more than offset by reducing the need for new wind and solar farms and other capacity. "Even though it takes a lot to build, it will pay for itself very quickly," Clack told me. Building more transmission lets us build less of something else—wind farms, solar farms, geothermal wells, hydroelectric dams, nuclear plants, and so on—since we will not waste their power in transmission bottlenecks. That can ease the transition away from fossil fuels and all the coal mines, fracking sites, ash ponds, leaks, and other impacts that come with them.[50]

But building new transmission lines is no easy task. Permitting decisions are made primarily by states, whose processes are inconsistent and uncoordinated. State and county officials and residents may object to lines that serve the greater good but don't convey much benefit locally. Interstate lines are especially difficult to build, since they require approvals from many jurisdictions.[51]

A chicken-and-egg problem stifles the growth of renewables and transmission: developers can't build wind and solar farms where there are no transmission lines, and utilities won't build transmission lines to empty spaces. "Transmission is fundamentally the limiting factor in tapping into the best resources," said Vanessa Tutos, director of governmental affairs at EDP Renewables.[52]

Texas partially solved its chicken-and-egg problem by designating certain windy areas as "competitive renewable energy zones" and directing utility companies to build transmission lines to them. Transmission investments paid off with savings from affordable wind generation, and they expanded opportunities for solar too. Wind power output more than doubled, while far less was curtailed by bottlenecks in transmission.[53]

Building lines across states has been harder. Clean Line Energy tried for a decade to build high-voltage lines connecting windy areas with more populated regions. I visited Clean Line's downtown Houston headquarters early in my work for this book, hoping to profile innovators who were helping to confront climate gridlock while making money too. "Who's making any money at this?" CEO Michael Skelly retorted. Sure enough, the company closed two years later, stymied by not-in-my-backyard opposition to power lines and by monopoly utilities warding off competition, in a saga chronicled by Russell Gold in *Superpower.*[54]

Rather than lines crossing many states, like Clean Line failed to build, or lines within a state, like Texas built, what's needed most is connectors to transmit power across the seams of neighboring grids. NREL and its partners conducted an Interconnections Seam Study to explore how such connectors could allow electricity to be decarbonized more reliably and affordably. The report, whose release was initially blocked by the Trump administration, showed that building interconnections could yield over twice as much savings as costs while helping wind and solar power reach more consumers.[55]

In his own analyses, Clack determined that just around 150 carefully located lines tying together existing grids would make it far more affordable to decarbonize electricity. "We don't need super-long transmission lines, but lots of smaller lines that can shuffle power around between regions," Clack told me. He recommends building the lines underground, to allay local concerns and protect the lines from wildfires and other hazards. Placing them along rail lines or highway corridors would minimize disruption and make power available to vehicle charging stations.[56]

Net-zero scenarios from both Princeton and the Zero Carbon Consortium would triple the nation's high-voltage transmission capacity by 2050, for an investment of roughly $2 trillion. Most of the lines would connect windy areas with regions of high demand, since wind is more geographically constrained than solar. Since transmission investments would be spread across decades and reduce fuel costs, they could more than pay for themselves. Researchers from the Massachusetts Institute of Technology found that coordinating transmission expansion on an interstate basis, rather than state by state, can dramatically reduce the cost of renewable electricity.[57]

To promote interstate coordination, Congress should pass legislation directing the Federal Energy Regulatory Commission (FERC) and partner agencies to develop a national strategy for transmission, as the House Select Committee on the Climate Crisis proposed in 2020. Such a strategy should not just consider immediate needs but also prepare for more wind and solar and growing demand from electrified vehicles, heating, and industry. A streamlined permitting process will be essential to expedite construction.[58]

Even without new legislation, the Energy Policy Act of 2005 authorizes DOE to designate national interest transmission corridors, where FERC can preempt state vetoes of new lines. It also authorizes DOE to partner with private entities to facilitate the construction of transmission lines. Although that authority was not enough to help Clean Line build a line from the Oklahoma panhandle to Tennessee, it could help other companies build the shorter connectors that are urgently needed. Still, executive branch actions under the 2005 act cannot match the comprehensive efforts that new legislation could unleash.[59]

Storage

Even with a diverse supply, expanded transmission, and demand flexibility, we will still have gaps and surpluses between supply and demand. That's where storage comes in. Storage is sometimes miscast as the "holy grail" for maintaining balance on a mostly wind and solar grid. Batteries are indeed getting cheaper. But we'll never build enough batteries to back up the grid. Batteries are costly to build and costly to operate, since energy is dissipated each time they are charged and discharged.[60]

Transmission moves power more efficiently. Complementary resources smooth out supply. Demand flexibility narrows gaps and surpluses between supply and demand. The more robustly we deploy complementary resources, transmission, and flexibility, the less storage we will need to build and the less often we will have to deploy it, reducing the overall costs of electricity.

Storage isn't just batteries. And it's not a one-size-fits-all commodity. Different options work better for providing power over different durations. Short-term storage maintains steady electrical frequency and smooths out supply across minutes or hours. Lithium-ion batteries work great for that, charging up with surplus power and discharging when power is scarce. Costs of lithium-ion batteries have fallen by 87 percent over the past decade and are expected to continue falling, making them the dominant option for short-term storage. Still, it would be cost-prohibitive to rely entirely on batteries to balance supply and demand. Building enough batteries to store just twelve hours of power nationally would cost over $2 trillion. "Lithium-ion batteries are an important complement to renewables, but they're no panacea," Jenkins told me.[61]

As the shares of electricity provided by variable wind and solar continue to grow, intermediate-term storage may become important. ARPA-E's DAYS (Duration Addition to ElectricitY Storage) program funds projects that develop and demonstrate technologies to store power over durations of ten to one hundred hours. That's long enough to complement solar power through the night or get through several days with cloudy skies, slow winds, or extreme weather conditions. DAYS-funded concepts include pumping water underground, heating up beds of magnesium-based particles, and building flow batteries that store energy in tanks of electrolytes.

Beyond intermediate storage, longer-duration options could be needed to supplement a mostly wind and solar supply through extended freezes, when sunshine is scarce, winds are slow, and heating demand peaks. It would also help during extended heat waves, when winds are slow and cooling demand peaks. Climate change is exacerbating summer extremes. Since different regions experience freezes and heat waves at different times, wheeling power around regional grids via enhanced transmission could slash storage requirements. Still, long-term storage can help grid operators manage extreme freezes and heat waves that drive up power demand and may be accompanied by slow winds across many states.[62]

Long-term storage has traditionally been provided by pumped hydropower—using surplus power to pump water up to a higher reservoir and then releasing it through a turbine during times of scarcity. Pumped hydropower can be built only on suitable terrain and with adequate supplies of flowing water. Very few sites have been built since the 1990s, although DOE's Hydropower Vision identified opportunities to add 36 gigawatts of new pumped hydropower capacity to the 22 gigawatts that exists today.[63]

Far greater capacity for long-term storage could come from storing hydrogen underground. As the world's lightest molecule, hydrogen (H_2) holds an exceptional amount of energy per unit of mass. Various uses and production methods for hydrogen are discussed in Chapter 6. Here, I'll focus on its potential for electricity storage.

Most hydrogen today is produced from fossil fuels and steam and used to make fertilizers and other chemicals. Called "grey hydrogen," its production is highly emissions-intensive and not a means for cleanly storing electricity.

"Green hydrogen" is instead produced by using clean electricity to split hydrogen from water in an electrolyzer. Doing so for electricity storage is nothing new. Large electrolyzers were built as early as the

1920s to make hydrogen from surplus hydropower from dams. As wind and solar farms proliferate, electrolyzers could make green hydrogen from their surplus output.[64]

Green hydrogen could be stored underground in formations such as salt caverns or depleted oil and gas reservoirs. Natural gas is stored today in this manner, which keeps it available throughout wintertime peaks in demand. Hydrogen could be stored underground for months with little leakage.

Stored hydrogen could be extracted whenever needed and used in fuel cells or gas turbines to generate electricity. Some gas turbines can already burn blends of hydrogen and natural gas, and new ones could be designed to run on hydrogen alone. Early deployments of flex-fuel turbines are planned for Utah, where growing percentages of green hydrogen from renewables will be blended with natural gas until turbines run completely on hydrogen. Some of the hydrogen will be stored in an underground salt dome, making it available when needed most.[65]

The big drawback of hydrogen storage is that its "round-trip efficiency"—how much energy is discharged compared to how much energy is put in—is only around 35 percent. By contrast, many batteries top 80 percent. Thus, for short-term purposes, we are better off using batteries. With adequate transmission, batteries plus some intermediate storage are enough to meet our needs while wind and solar are scaling up over the next decade or two. Only when wind and solar percentages grow very large will long-term storage become important, especially for getting through the depths of winter as more heating is electrified. As energy modeler Tom Brown of the University of Karlsruhe put it, "A wild expansion of electrolysis [for hydrogen] only makes sense when we have renewable power coming out of our ears. Until then, the priority is to build renewables faster, and start scaling up electrolyzer manufacturing so it's down the learning curve when we need it."[66]

Aiming High

Pulling together all the tools of complementary resources, flexible demand, transmission, and storage alongside wind and solar, we can certainly decarbonize the power supply very deeply while making it more affordable, reliable, and resilient. Numerous studies project the share of GDP spent on energy could continue to decline even as we decarbonize. But should we aim for 100 percent clean electricity? Or is getting close good enough?

The clarity of a 100 percent target cannot be dismissed. "We don't go 98.5 percent of the way to the Moon," said Mark Jacobson, a professor at Stanford University who has modeled scenarios for 100 percent renewable electricity. Like President Kennedy's call to go to the Moon, a 100 percent target serves as a clarion call for action. It makes clear that fossil fuel plants and the pipelines and coal mines that serve them have no place in the grid of the future. Knowing that, we can begin planning for environmental remediation and just economic transitions for communities near retiring facilities, even as clean energy creates more jobs overall. Developers and deployers of clean electricity technologies can redouble their efforts, counting on a booming market for renewables, transmission, storage, and demand-flexing devices. Grid operators, public utility commissioners, utility executives, and others can plan for the transition as well.[67]

The clarity of 100 percent targets has been winning out politically. By 2020, six states plus the District of Columbia and Puerto Rico had already committed to 100 percent carbon-free electricity by 2050 or sooner, and eight others had set 100 percent as a goal. The Green New Deal proposed a national target of 2030, and President Joe Biden's climate plan calls for 100 percent clean electricity by 2035. Ironically, all these policies and proposals were set after President Trump took office with a vow to boost coal.[68]

Despite the political appeal, mandating 100 percent clean electricity too soon would be self-defeating. Since carbon dioxide accumulates and coal pollution is deadly, eliminating coal and scaling up clean sources as quickly as possible can be as important as ultimately scaling them to 100 percent. Clean electricity is not an isolated pillar of clean energy. Instead, it is profoundly interconnected with efficiency and electrification. Efficiency depends upon electric cars and heat pumps replacing gasoline cars and gas furnaces. Electrification depends on electricity being affordable and reliable enough to replace other fuels wherever practical. Even some negative emissions technologies, like capturing carbon dioxide from the air, depend on abundant and affordable electricity. If demands for purity make electricity less affordable or reliable for the sake of squeezing out every last ounce of emissions, that outcome would undermine the construction of the other pillars.

Mandating 100 percent clean electricity and eliminating gas would constrict our options to satisfy demand at the toughest times, when gas plants can be dispatched quickly to avert shortfalls. Without some flexibility from gas, we might need to overbuild and rarely use something else—extra wind and solar farms, geothermal or nuclear plants, transmission lines, or storage—to keep the lights on when clean supply falls short of demand, especially as more electrified heating intensifies wintertime peaks in demand. "The more you electrify, the more it becomes nonlinear how much you have to overbuild," said Michael Webber, a professor of mechanical engineering at the University of Texas at Austin. The Zero Carbon Consortium's net-zero strategy would keep some gas plants on hand, deploying them sparingly to satisfy peak demand. Blending in hydrogen or biogas could make those gas plants cleaner, but the paucity of use is what matters most. If this route is taken, policies must make it financially viable to retire some gas plants while keeping others maintained and fully weatherized to be ready for dispatches that come as rarely as possible.[69]

Recent research from the University of California at Berkeley found that a 2035 deadline for 100 percent clean electricity nationwide would leave insufficient time for new generation and storage technologies to scale up to balance wind and solar year-round. However, it found that aggressive policies could achieve a target of 90 percent by 2035 while keeping electricity prices below their current levels. Such an ambitious yet achievable target would yield enormous benefits while setting the stage for widespread electrification and deeper decarbonization of electricity over the subsequent decade.[70]

America's Role Abroad

How could America's pursuit of clean electricity influence parallel pursuits abroad? In some ways, it will not. Very little electricity crosses our borders. Dozens of countries already generate electricity far more cleanly than the United States, with non-fossil sources providing the bulk of their supply. Their leading sources of clean electricity vary widely, from the nuclear power of France to the wind power of Denmark to the hydropower of Norway and much of Latin America. With a far larger grid and far more coal to displace, China has been adding wind and solar power several times more quickly than the United States, so it will be a stronger driver of learning curves from deployment.[71]

Furthermore, the clean electricity mix that emerges here will not necessarily be replicable abroad. The United States is blessed with a uniquely diverse array of resources. Its central states have stronger winds than just about anywhere in southern Asia or sub-Saharan Africa. The northeast and northwest coasts feature strong offshore winds. The desert Southwest gets more sunshine than anywhere in Europe. Even the upper Midwest gets more sunshine than most of Germany or Japan, the countries that rank just behind us in solar installations. Western states offer abundant untapped opportunities for geothermal power. Various regions sit atop geology that is well-suited for storing carbon dioxide or hydrogen. Meanwhile, legacy fleets of hydroelectric dams and nuclear power plants provide firm output to balance variable renewable electricity.

Where the United States can lead is in driving a technology push toward clean electricity innovation. The United States ranks second only to China in renewable energy patents, producing more than all other developing countries and the European Union combined. Our national laboratories and universities, plus an unmatched wealth of public and private funders, provide fertile ground for innovation. A handful of those funders, such as ARPA-E and Breakthrough Energy Ventures, specifically seek out high-risk, high-reward endeavors that could lead to technology breakthroughs.[72]

Several aspects of clean electricity are especially suited for U.S. leadership. As the country that pioneered fracking and horizontal drilling, and with abundant geothermal resources in western states, the United States could pioneer enhanced geothermal technologies. Success with those technologies would make geothermal energy accessible in far more locations, and it would leverage the expertise of

American oil companies as demand for oil declines. Those companies can also lead the way toward capturing carbon dioxide and storing it in oil and gas reservoirs and other geological formations. Meanwhile, developers of oil and gas pipelines can redirect their efforts to retrofitting those pipelines or building new ones to transport hydrogen and captured carbon dioxide. Finally, Silicon Valley can drive innovations that make efficiency and flexible demand more appealing to consumers. In all these areas, American innovations can expand opportunities and drive down costs for decarbonizing electricity in other countries.

Okay, so we've cleaned up electricity. That raises two questions. What can run on that clean electricity? And how can we cleanly fuel everything else?

Electricity already powers nearly 40 percent of U.S. energy needs, but its use is spread unevenly across the economy, resulting in a jagged frontier between what's electrified and what's not (Figure 8). Even as electricity has gotten cleaner, that frontier has barely budged. Transportation still uses hardly any electricity, driven mostly by liquid fuels. By contrast, homes and businesses already get most of their energy from electricity, using it to power air conditioners, refrigerators, lights, and appliances. Space and water heating are the main exceptions, as gas, propane, and fuel oil compete with electricity for market share. Meanwhile, many industrial processes, such as the manufacturing of chemicals, paper, cement, and steel, remain largely unelectrified. Mining and refining also consume huge amounts of fuel, but that usage would shrink if decarbonizing other sectors lets us mine, refine, and transport less fossil fuels.

This means that any battle to extend the electrified frontier into fossil-fueled terrain must be waged on three main fronts: transportation, heating, and industry. Extending that frontier will enable increasingly clean electricity to power more of the economy. However, as we'll see, those extensions will also transform the challenge of decarbonizing electricity itself. Meanwhile, hydrogen, advanced biofuels, or other clean fuels will be needed to decarbonize whatever is not electrified. Trillions of dollars and billions of tons of emissions are at stake in getting it right.

Decarbonizing Cars

Light-duty vehicles such as cars, pickup trucks, and sport utility vehicles emit more carbon dioxide than all other transportation combined in the United States. COVID-19 lockdowns aside, policies have been

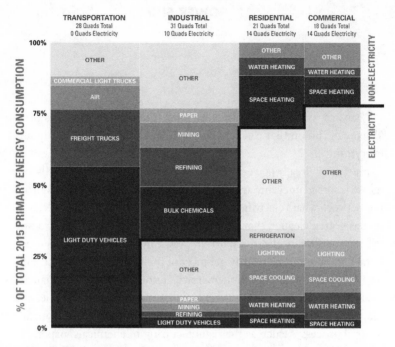

Figure 8. U.S. energy use in 2015 (1 Quad = 1 quadrillion Btu). Activities below the bold line were powered by electricity. (Image courtesy of NREL Electrification Futures Study, nrel.gov/efs)

largely ineffective in curbing the inexorable growth in how far we drive. Americans travel nearly 90 percent of their mileage in privately owned vehicles that average just 1.5 passengers, even as the size of vehicles has grown. Tweaks to urban design such as bike lanes, rezoning, and denser development can only do so much to reduce driving within cities already built around automobiles, and local improvements don't affect travel between and beyond cities. Thus, much as I would love to see cities become more walkable, bikeable, and accessible via public transit, making them all-around better places to live, emissions cuts for the foreseeable future will come mainly from reducing emissions per mile rather than reducing miles traveled.[1]

Efficiency can help. Hybrid cars such as the Toyota Prius achieve twice the fuel economy of traditional vehicles. Eventually, though, efficiency can cut gasoline use only so much, and there is no way to capture the tailpipe emissions from burning it. Burning any fossil fuel generates more than twice its weight in carbon dioxide, as its carbon bonds with oxygen from the air.

Deep decarbonization will require powering vehicles without burning fossil fuels. Until recently, it was not clear that electricity was the best way to do so. "A decade ago, we thought it was really a jump ball between electric vehicles and fuel cell vehicles," said Amber Mahone of Energy and Environmental Economics. Hydrogen fuel cells use just hydrogen and air to make energy. If the hydrogen is produced cleanly—a big if, since most hydrogen today is produced from natural gas and steam—fuel cells provide a clean way to power a car, emitting only water vapor from the tailpipe. "With hydrogen, you can store more energy more compactly and at lower weights than you can with batteries," said Joan Ogden of the University of California at Davis. Hydrogen can refuel a vehicle nearly as quickly as gasoline or diesel and far faster than a battery can be recharged.[2]

Despite these advantages, market conditions have shifted decisively in favor of electric cars. "Now, electrification of light-duty vehicles is basically a no-brainer," Mahone told me. Electric vehicle sales are outpacing fuel cells by a factor of 100. With growing economies of scale, electric vehicle costs have fallen far below the cost of comparable fuel cell ones. Price gaps could widen, as car manufacturers spend most of their clean vehicle R&D budgets on electrics. Industry and government at the state and local levels are collaborating to build out charging infrastructure and promote electric vehicles, while support for fuel cells remains scant. Just forty-four public hydrogen refueling stations were operating in the United States in 2020.[3]

Most pivotal to the rise of electric cars has been the falling costs of their most expensive component, lithium-ion batteries. Battery costs dropped by 89 percent in real terms from 2010 to 2020, bringing the sticker prices of electric cars near parity with gasoline ones. Beyond upfront costs, electric cars are cheaper to operate and maintain than gasoline or diesel ones, thanks to fuel savings and fewer moving parts that can degrade. For example, a teardown of an electric Chevy Bolt and a gasoline VW Golf found that the Bolt had just 35 moving and wearing parts, compared with 167 in the Golf.[4]

Despite falling costs, electric vehicles face barriers in the marketplace. Car dealers tend to be reluctant to sell electric cars, which are less familiar and which may yield fewer lucrative trips to their service shops. Car buyers focus mostly on sticker prices, discounting fuel and maintenance savings that accrue over time. Buyers may also be deterred by "range anxiety"—a fear of running out of charge while driving. Such anxiety took root when early electric vehicles failed to live up to even their limited promised ranges. When Nissan introduced the Leaf as the first mass-market, fully electric vehicle in 2011, it claimed a range of 175 kilometers (109 miles). Testing by the EPA found its actual range was just 117 kilometers (73 miles), and drivers complained of even shorter ranges during cold weather.[5]

Now that many electric vehicle models can reliably travel more than three hundred kilometers between charges, range anxiety has given way to a new concern—"charging time trauma." Ranges of a few hundred kilometers far exceed most people's daily commutes and would easily suffice for road trips if drivers could refuel quickly. But recharging an electric car isn't as quick and easy as pulling in to the nearest gas station. As of 2020, the United States had just 26,000 public electric vehicle chargers, compared with more than 100,000 gasoline stations. Whereas many gas stations offer multiple pumps for refueling in the time it takes to buy a soda, only a small fraction of charging stations offer fast charging, many of them dedicated solely to Teslas. The rest take several hours to recharge a car, and even fast chargers take half an hour. "You can never charge batteries as fast as you can fill a vehicle with fuel—it's a fundamental limitation," said Jack Brouwer, director of the National Fuel Cell Research Center at the University of California, Irvine.[6]

Even if recharging an electric car never becomes as fast as refueling with a liquid fuel, drivers' concerns can be assuaged if enough public chargers are available to complement charging at home or work. Analysts from NREL estimated that 1,850 fast chargers and 40,000 "Level 2" chargers are needed for each million electric vehicles. Extrapolating to net-zero scenarios, Princeton University researchers estimated that more than 2 million public chargers (including over 100,000 fast ones) would be needed to power 50 million electric cars on U.S. roads by 2030 (up from 5 million in 2020), and 16 million public chargers to power a mostly electric U.S. fleet by 2050. Daunting as those numbers may seem, building such a charging infrastructure would cost just $7 billion in the 2020s and $20–25 billion in each subsequent decade, the study estimated, money that could be more than recouped from drivers.[7]

Building a sufficient public charging infrastructure is important not just for overcoming range anxiety and charging time trauma, but also for addressing equity concerns. A homeowner with a garage and a thousand dollars to spare can readily install a charger. A renter may not have that luxury. Even as some apartment complexes and office buildings begin installing a handful of chargers, the numbers are nowhere near what would be needed if most cars become electric.

Equity concerns also arise from the fact that, for now, electric cars are mostly bought by well-off drivers who can afford to keep an extra car on hand or rent one for longer trips. Higher upfront costs deter low-income buyers, who cannot afford to wait for fuel and maintenance savings down the road. Tax credits apply only to new cars and come through a rebate of income taxes paid that year, making them unavailable to used-car buyers or to new-car buyers with little taxable income. Better designed incentives and lower sticker prices will be needed to address those inequities.

Decarbonizing Buses and Trucks

Whereas batteries provide the best means for decarbonizing cars and light trucks, they face challenges in heavier vehicles. Lithium-ion batteries carry just 2 percent as much energy per kilogram as diesel. Solid-state batteries could improve energy density and reduce flammability, but they are for now far costlier than lithium-ion ones. Even with improved batteries and efficient electric motors, electric trucks would weigh many tons more than their diesel competitors.[8]

Heavy batteries are not a deal-breaker for local vehicles such as delivery trucks, transit buses, and school buses, since they can return to a depot to recharge. That limits their battery needs. Electrics already provide lifetime savings for local trucks and buses. Fleet owners are more attuned than typical consumers to valuing ongoing savings. And because electric buses and delivery vehicles are driven mostly on busy urban streets, their quieter and exhaust-free operation provides welcome relief to riders and pedestrians weary of the fumes and noise of diesel engines.

Fumes and noise are not mere nuisances. In fact, the California Air Resources Board estimates that more than a thousand cardiovascular deaths and 70 percent of the cancer risk from toxic air contaminants in the state come from diesel exhaust. Diesel engine noise contributes to hearing loss, sleep disturbance, and cardiovascular disease. The societal value of avoiding that noise can be even larger than the climate benefits of reducing emissions, studies have shown. Thus, even though light-duty vehicles have been called a "no-brainer" for electrification, it's the local diesels such as urban buses that deserve top priority. And since diesel buses operate so much and with such poor fuel economy, replacing each one can save huge amounts of fuel and emissions. In fact, in 2019 the electric buses in Chinese cities alone were saving more fuel than all the electric sedans in the world combined.[9]

All these considerations should make electrification the obvious choice for local trucks and buses. That's not yet the case for long-haul trucking. Unlike buses that can recharge locally, long-haul trucks are typically expected to travel long distances between refueling stops. The lithium-ion batteries needed to achieve the ranges that truckers demand would take up far more space than diesel tanks and eat up much of the truck's highway weight limit. That would leave less space and weight available for carrying payload. More energy-dense batteries under development would ease but not eliminate those burdens. Such large batteries would take hours to charge with conventional chargers or would require new chargers with enormous capacity, comparable to the output of several acres of solar panels. None of these challenges are insurmountable, but they suggest that hydrogen fuel cells deserve a closer look.[10]

Hydrogen fuel cells combine hydrogen with oxygen to generate electricity to power a vehicle, releasing just water vapor as exhaust.

Hydrogen's advantages of faster refueling and lighter weight are especially valuable for long-haul trucking. Truckers typically refuel along highways rather than in neighborhoods, so fewer fueling stations are needed.

Hydrogen fuel cell heavy-duty trucks were first introduced in 2009 for local use in California ports. Now some of the biggest manufacturers in the world, such as Hyundai, Daimler, Volvo, and Toyota, have announced plans to build fuel cell trucks, but most are hedging their bets and designing electric ones too. With technology learning rates and economies of scale, some analysts expect hydrogen fuel cell trucks to near purchase price parity with diesel soon.[11]

The snag, at least for now, is that hydrogen fuel is costlier per mile than diesel, so upfront price gaps can't be recouped. That is the case even with "grey hydrogen," whose production from fossil fuels negates much of its emissions savings. "Green hydrogen," made by splitting water with clean electricity, will be even pricier until electrolyzers and electricity become much cheaper. Thus, some analysts expect batteries to outcompete fuel cells for lifetime costs. "I would be very disappointed if less than 50 percent of new trucks by 2030 are electric, and I would not be surprised if it's 90 percent," said Auke Hoekstra, an expert on electric vehicles at Eindhoven University of Technology. Others see hydrogen fuel cells having the edge. "In our models, we think fuel cells will dominate long-haul trucks," said Marshall Miller of the University of California at Davis, arguing that truckers would shy away from heavy batteries that reduce range and payload capacity. "I think something would have to change for battery electric vehicles not to dominate in every area but long-haul, and for fuel cells not to dominate for long-haul."[12]

As the relative prospects of battery and fuel cell trucking remain disputed, the prospect of both gaining substantial market share raises thorny challenges. The infrastructure needed to provide either ultrafast charging to electric trucks or high-capacity hydrogen refueling to fuel cell ones would cost a few million dollars per truck stop. Providing both would of course cost more and require both power lines and hydrogen delivery systems or on-site production. Thus, while the race to decarbonize trucking is close for now, if either electricity or hydrogen pulls into a lead, its growing infrastructure and economies of scale could propel it onward.[13]

With electricity already winning the race for light-duty vehicles, it is hydrogen that runs the greater risk of being left behind. Green hydrogen is still at least a decade away from widespread availability. The transmission of electricity is already ubiquitous, but hydrogen supply is not. "You get to a chicken and egg problem with fuel cell trucks and fuel cell vehicles of any kind—how do you support the build-out of the hydrogen fueling network," Mahone said. There is no guarantee that such support will emerge. "Hydrogen needs a compelling advantage, and I don't see it," Hoekstra told me. But if industry or government does support an initial construction of hydrogen infrastructure and investments in fuel cell trucks, further growth could become self-sustaining. "As you get more stations out there, the costs of hydrogen will come down with scale, and you'll get to the point where hydrogen is competitive with gasoline and diesel per mile and the system pays for itself," Ogden told me.[14]

Decarbonizing Ships and Planes

Ships and planes will be far more difficult to decarbonize than cars and trucks. Shipping and aviation are each responsible for just 2 percent of carbon dioxide emissions globally, but those shares are expected to expand as other sources are decarbonized and travel increases. To travel long distances without refueling, planes and ships need tremendous amounts of energy. The poor energy density that makes batteries challenging for use in long-haul trucks makes them virtually inconceivable for long-distance ships and planes. It is therefore unlikely that we will see electricity powering much more than ferries or small planes anytime soon. Oil-based fuels are so attractive for ships and planes because they pack so much energy per volume and weight. Similar hydrocarbon fuels ready for use by existing ships and planes could be produced from biomass or synthesized from hydrogen and captured carbon. Since most ships and planes operate for decades, those "drop-in" options will be crucial for cutting emissions before ships and planes are replaced.[15]

Eventually, hydrogen or synthetic fuels produced from hydrogen and captured carbon could power a new generation of ships and planes. Hydrogen fuel cells have already been demonstrated in ferries and submarines. Small planes have been designed to use liquid hydrogen in fuel cells or burn it in engines. Longer-distance travel requires denser fuels, such as the synthetics that will be discussed below.[16]

Decarbonizing Heating and Cooking

Apart from transportation and power plants, more fossil fuel is burned to heat air and water in our homes and businesses than for any other purpose. Most of that burning uses natural gas, which heats the air and water in nearly half of U.S. homes and most businesses. Fuel oil and propane are burned in far fewer homes, with most fuel oil used in New England and most propane in rural regions. That leaves electricity heating the air in 39 percent of U.S. homes (up from just 2 percent in 1960) and water in about half, with far smaller shares in businesses. For heating as for transportation, then, two intertwined questions arise: how much more to electrify, and how to decarbonize the rest.[17]

As in transportation, efficiency can help a lot, even without electrification. Programmable thermostats, insulation, efficient windows, and better maintained ventilation systems can slash heating and cooling needs. For the dwindling number of homes that still use fuel oil, switching to gas furnaces can reduce emissions by a third. For homes that already use gas furnaces and water heaters, replacing them with gas-fired heat pumps could cut fuel use for space and water heating by half, though such devices have yet to become widely available.[18]

Those efficiency-improving steps would be sufficient if our aim was merely to cut emissions moderately and quickly. Until electricity becomes cleaner, gas heating can have a smaller carbon footprint than electricity in some regions. Burning coal or gas in a faraway power plant and transmitting it to a home is a lot less efficient than burning gas on-site, if the electricity is used in old-fashioned "resistance" space and water heaters that, like toasters and hair dryers, can provide no more heat than the electricity they consume.[19]

But to deeply decarbonize heating and cooking, we will need to electrify them. Only transitioning away from fuel oil and gas can eliminate the emissions associated with distributing them and the indoor air pollution that comes from burning them in homes. Heating and cooking with fossil fuels are the leading sources of carbon monoxide and nitrogen dioxide air pollution in homes. Methane, the main component of natural gas, leaks out every step of the way from drilling sites to pipelines to water heaters and furnaces and stovetops. A 2020 study found that local gas distribution systems leak five times as much methane as EPA had estimated, making them responsible for 8 percent of all methane emissions nationwide. Since gas pipes must stay constantly pressurized, averting those leaks will require not just reducing gas use but discontinuing it wherever possible. "If we want a climate-safe world, we're going to need to phase out most if not all gas burning," said Bruce Nilles, executive director of Climate Imperative.[20]

Heat pumps can provide heat several times more efficiently than furnaces, and they can be designed to provide air conditioning, space heating, and water heating year-round. Although their efficiency advantages are most pronounced in mild weather, the latest heat pumps can operate efficiently at temperatures of $-15°C$ ($5°F$) or less. Hybrid systems use electrical resistance to boost heat output at the coldest times. Ground-source systems circulate their fluids through tubing underground, where temperatures stay stable year-round even as air temperatures swing. That offers superior performance and energy savings, but this technology isn't widely available yet.[21]

Converting old buildings from fossil furnaces to electric heat pumps can require upgrades to electrical systems, along with the purchase and installation of the heating systems themselves. Those upfront costs may be tough to recoup with fuel savings, if natural gas remains cheap. Even homeowners willing to incur those costs for the sake of the environment may struggle to find contractors experienced in electrifying old homes. Climate activist Justin Guay chronicled his months of frustration as one contractor after another told him that electrification would be impossible or outlandishly expensive for his 1950s-era California home. "I highly doubt any but the most committed would ever undergo all of it," he told me. Incentives will be needed to spur homeowners to act and to train contractors in home electrification.[22]

Policies will also be needed to ensure a just and equitable transition away from local gas systems. As homes and businesses become more efficient or disconnect from gas by electrifying, the fixed costs of operating and maintaining aging gas distribution systems will be split across fewer and fewer customers using ever less gas. Poorer homeowners and tenants may face rising gas bills but lack the means to electrify and disconnect from gas. Careful planning is needed to help them electrify their homes.

Transitions from gas to electricity may vary by region. In the coldest regions, continuing to heat with gas could avoid spikes in electricity demand during cold snaps, when solar energy is scarce and electric heat pumps are not as efficient. Cold fronts arrive far faster than heat waves, making winter power demand especially spiky. "It might make sense to keep gas distribution systems in northern parts of the Northeast and Midwest, and transition to all electric in warmer climates," Mahone said. Other experts argue that heating could be electrified even in the coldest regions by pairing heat pumps, ground-sourced if possible, with electric resistance backup. "Keeping a gas distribution system just to meet peak needs on the coldest days is not likely to be economical," said Mike Henchen of the Rocky Mountain Institute.[23]

In warmer regions, what should be sought is a strategic retreat from local gas distribution altogether. Setting a specific date, perhaps around 2045 in most regions, for shutting off local gas distribution will give customers time to switch to electricity when their gas heaters need replacement; setting a clear time frame will also prevent gas utilities from building assets that will soon be stranded. For older homes, incentives could help low-income homeowners disconnect from gas and switch to electric heat pumps and stoves, cutting their bills and improving the air they breathe. New homes should be built without gas hookups at all, enabling them to be fully electrified from the start with the most efficient devices possible and sparing the costs of expanding gas infrastructure.

With its goal to reach net zero by 2045, California has become a testbed for transitioning away from local gas. Dozens of cities have followed a decision made by Berkeley in 2019 to ban new gas hookups, and several environmental groups and public-private partnerships have issued blueprints for equitable and affordable transitions from gas to electricity. Although gas-only utilities have vigorously opposed bans on gas for new buildings, persuading legislatures in some states to prohibit local bans, California's largest combined gas and electric utility supported the ban in Berkeley.[24]

The biggest obstacle to an orderly transition away from residential gas may come from one of its smallest uses—cooking. Less than 4 percent of gas use in homes is for cooking, which is outranked even by hot tubs. Yet gas cooking has a unique hold on the imagination, keeping consumers hooked on maintaining gas service. "Most people don't care how their water is heated or how their heater works, but the Viking stove in their kitchen, people have this visceral emotional attachment to," Michael Colvin of the Environmental Defense Fund explained to me.[25]

In fact, many chefs and consumers who have tried electric induction stoves prefer them. Induction stoves work by using electromagnets to transfer heat directly to magnetic cookware rather than creating a flame or heating the cooktop itself. That lets them heat food and boil water nearly twice as fast as gas, saving time and slashing energy use. It also means that induction cooktops don't stay hot after their use, minimizing the risk of accidental burns. Prices of induction stoves have been falling, and some models have earned nearly perfect scores from *Consumer Reports*. Traditional electric cooking is less efficient, but it enables electrification with a more familiar technology.[26]

The challenge is overcoming misperceptions. The notion that gas offers superior cooking has been peddled since the 1930s, when the American Gas Association (AGA) introduced the advertising slogan "Now you're cooking with gas." An AGA executive planted the phrase with writers for the comedian Bob Hope, and the phrase was soon picked up by Jack Benny and even Daffy Duck. AGA continues to employ the phrase as a hashtag in promotional videos. "Cooking is the hill that the gas industry wants to fight on," said Nilles. "They'll say, 'Do you want the government to take away your gas stove that makes you a great chef?' "[27]

What the industry does not mention is that gas cooking is a leading source of air pollution in homes, including levels of nitrogen dioxide that can far exceed EPA limits for outdoor air. A meta-analysis found that gas cooking in the home increases the risk of asthma in children by 32 percent. Gas appliances have also been associated with impaired cognition and attention in preschoolers. This means both health and climate are at stake in the transition away from gas. "A few decades from now, people will say 'What were we thinking burning fossil fuels in our homes,' " said Jacob Corvidae of the Rocky Mountain Institute.[28]

Scheduling end dates for local gas distribution will not be popular. The gas industry aggressively promotes its use, sponsoring social media influencers to promote the "Natural Gas Genius lifestyle" and funding astroturf groups to drum up local opposition to gas bans. But firm policies beat the alternative. Without them, gas bills will spiral higher as a dwindling number of customers bear the costs of aging distribution systems that continue to leak methane. By contrast, electric heat pumps will likely continue becoming cheaper. "Economics is going to make electrification make more and more sense," Colvin said. But without policy augmenting economics, burdens will fall most heavily on the vulnerable, money will be wasted on soon-to-be-stranded assets, and methane leaks will persist. Sensible policies, held firm in the face of genuine and drummed-up opposition, can ensure a more orderly, equitable, and just transition.[29]

Decarbonizing Industry

Industries burn fossil fuels to provide most of the heat, steam, and electricity required to manufacture chemicals, paper, steel, concrete, and other products. Since the needs of different industries vary widely, there is no one-size-fits-all answer to decarbonizing them. However, a few key themes stand out.

First, it takes fossil energy to make fossil energy. About 12 percent of all energy use goes to mining, extracting, refining, and transporting fossil fuels themselves. Meanwhile, nearly one-third of domestic rail tonnage is coal, and about 40 percent of international maritime shipping is devoted to transporting fossil fuels. Therefore, any cuts in fossil fuel use beget more cuts in fossil fuel use.[30]

Second, various manufacturing processes require huge amounts of heat, at specific temperatures for each process. Globally, producing all that heat results in about 10 percent of all greenhouse gas emissions—more than cars and planes combined. Manufacturers of steel, cement, and glass all require temperatures well above 1,000°C. Other manufacturing requires heat at lower temperatures, but still lots of it. For now, the vast majority of all this heat comes from burning fossil fuels on-site.[31]

Electricity can replace fossil fuels for even the highest-temperature industrial heat needs. For example, electric arc furnaces can supply the 2,200°C temperatures needed for steel manufacturing. Across a wide range of applications, electricity enables factory operators to control heat levels more precisely and thereby reduce fossil fuels and the wastes that come from burning them. However, switching to electric heat requires redesigning systems that had been engineered around fossil fuel burning.[32]

Other low-carbon sources of heat include wood scraps, which are widely used at pulp and paper mills. Nuclear, geothermal, and solar thermal power can supply heat or steam along with electricity to other factories. The California start-up Heliogen, backed by Bill Gates and others, is piloting the use of sunlight-reflecting mirrors to generate heat above 1,500°C, hot enough to manufacture cement or chemicals.

Hydrogen provides a tantalizing option because it can be produced almost anywhere and burned at very high temperatures or serve as a chemical feedstock. For example, hydrogen could replace natural gas as a feedstock for ammonia-based fertilizers. The trouble is cost. "Hydrogen is currently more expensive than natural gas and will require lower-cost electrolyzers and renewable electricity to compete," said Mark Ruth of NREL. DOE in 2021 launched Hydrogen Shot with an aim of reducing the cost of clean hydrogen 80 percent to $1 per kilogram within a decade.[33]

Low-carbon heat could also come from fossil fuels with carbon capture. Ethanol, ammonia, and cement manufacturing all yield far more concentrated streams of carbon dioxide than power plants, simplifying carbon capture. New pipelines would be needed to move carbon dioxide from those factories to geological repositories where it could be stored. "If we had the pipelines, we would triple or quadruple the number of projects immediately," Julio Friedmann, an expert on carbon capture at Columbia University, told me.[34]

Unlike electric cars and heat pumps, which may soon outperform their fossil rivals, decarbonizing industrial heat is likely to be costly and complex. For now, electricity, hydrogen, and carbon capture cost at least twice as much as burning natural gas for industrial heat. Manufacturers facing razor-thin margins for commodities such as steel and cement will not be able to bear the costs of decarbonization unless policy incentivizes or demands it. Climate clubs, border adjustment tariffs, clean material standards, or other policies must create an international playing field that favors clean manufacturing methods.[35]

Role of Electrification in Decarbonizing Electricity

As vehicles, heating, and industries increasingly electrify, their electrification will reshape the challenge of decarbonizing electricity itself. Whether electrification helps or hinders that undertaking will depend on how electrification is done. "You could electrify intelligently, or you could do it in a more costly manner, or in a way that reduces the environmental benefits of electrification," said Trieu Mai, who leads electrification research at NREL.[36]

First, some historical context. Electricity demand in the United States grew roughly 8 percent per year through the 1950s and 1960s as industry boomed. It continued to grow 2–4 percent per year until the turn of the century, as air conditioning and appliances proliferated in ever larger homes and businesses. Since then, growth has ground to a halt. Even before the COVID pandemic, electricity demand in 2019 was slightly less than it had been in 2007. Industries became more efficient and moved some operations overseas. Tougher efficiency standards for air conditioners, heaters, and appliances held residential and commercial electricity demand in check. Efficiency and slow population growth could keep electricity demand flat, except for demand from newly electrified products and processes. In NREL's most ambitious electrification scenario, with most vehicles and heating and some industries electrified by 2050, total electricity demand would grow just 1.6 percent per year, far slower than its historical pace. Growth would be faster in some of Princeton's net-zero strategies but still nowhere near the pace of last century. Thus, electrification will spark substantial but not unprecedented growth in electricity demand, while shifting when and where it's needed.[37]

Electrifying transportation would drive most of that growth and would pose distinct challenges. Most charging at home and work is done with Level 2 chargers, which consume nearly 20 kilowatts of power, comparable to the power use of ten homes. They can recharge a car in a few hours. Fast chargers need up to 350 kilowatts to recharge a car in fifteen minutes. Megachargers for big rig trucks could draw more than 1,000 kilowatts, comparable to an entire neighborhood. This would mean parking garages, truck stops, and fleet charging depots could all become huge consumers of electricity, with wide volatility in use rates. Power distribution systems must be scaled up to handle that demand.[38]

Charging will likely peak at offices in the morning when commuters arrive and at homes in the evenings when they return from work. Unfortunately, sunshine is weak at both of those times. Demand at service stations could peak in summer for family road trips and in winter when cold weather impairs battery performance. That could coincide with customary summer and winter peaks in demand from other sectors. As a result, vehicle charging poses both daily and seasonal challenges, especially as solar power grows and more heating is electrified.

To mitigate these difficulties, homes and offices can deploy "smart" chargers that schedule charging to occur when clean power is abundant. Plugging in a car overnight or throughout a workday provides plenty of time for such scheduling. Service stations and truck stops could vary the pricing displayed on their towering signs to attract customers when power is cheap.

Electrifying heating poses its own challenges. Even with efficient heat pumps, electrified heating would intensify winter peaks in electricity demand, when solar output is scarce. Like smart car chargers, smart water heaters can mitigate peaks by modulating when they heat water in the tank. Smart thermostats could pre-heat or pre-cool rooms when power is most abundant. An intriguing feature of heat pumps is that they operate most efficiently when temperatures are mild—so pre-cooling a well-insulated room on a mild summer morning could be a smart way to reduce demand during hot afternoons.[39]

Industrial electrification may pose fewer timing challenges. Factory managers are adept at modulating power use to avoid peak prices, so long as markets reward flexibility via variable prices or other means. Still, growth in industrial power demand, much of it for continuous processes, will heighten the importance of reliable power supply. After Winter Storm Uri triggered blackouts across Texas in 2021, some petrochemical plants struggled for weeks to ramp back up their production processes, straining supplies of some chemicals.[40]

Hydrogen

As we've seen, hydrogen has many potential uses, including electricity storage, trucking, chemical production, and industrial heat—which means options for producing and distributing hydrogen deserve a closer look. "Hydrogen is hard to make, hard to move, and hard to store, but once you've got it, it is a brilliant ingredient," Webber told me.[41]

"Grey hydrogen," made from methane and steam, dominates the market today. It is used mainly to produce fertilizers and other chemicals. Despite such narrow use, grey hydrogen production consumes 6 percent of natural gas and 2 percent of coal globally.[42]

"Blue hydrogen" is also made from methane and steam, but with the carbon dioxide captured. Syzygy Plasmonics is piloting a related approach that uses light and catalysts to convert methane to hydrogen, reducing energy use (disclosure: Syzygy funded a research project by my group). Either approach averts carbon dioxide emissions but requires obtaining methane, sending it through leaky pipelines, and finding somewhere to store the carbon dioxide. Critics argue that those steps keep blue hydrogen from being a truly clean option like "green hydrogen," produced by splitting water with clean electricity. However, boosters see blue hydrogen as a bridge to broader uses of hydrogen. "You can make blue hydrogen today at the largest possible scale you want at a fraction of the cost of the most attractive green hydrogen options," said Friedmann, the carbon capture expert at Columbia University. As of 2020, green hydrogen cost 1.3 to 5.5 times as much to produce as blue hydrogen, which itself was roughly 50 percent costlier than grey hydrogen, according to Goldman Sachs.[43]

"Green hydrogen" from clean electricity electrolyzing water could close those gaps if clean electricity and electrolyzers become cheaper or if natural gas becomes costlier or taxed. "With reasonable assumptions for cost curves for wind, solar, and electrolysis, we are finding that if you deploy it at scale, then renewable hydrogen can get pretty cheap," said Brouwer, the hydrogen expert at the University of California, Irvine. "Eventually, if hydrogen is going to become a major energy vector, the vast majority of it must come from sun and wind power." Plunging solar costs could make that vision a reality. "The next generation of solar technologies could be extraordinarily cheap, which would unlock a range of applications like green hydrogen," solar expert Sivaram told me.[44]

Green hydrogen could be produced anywhere on the power grid by electrolyzers of any size. By contrast, blue hydrogen production requires separate pipelines to bring in the natural gas, deliver the hydrogen, and dispose of the carbon dioxide. Thus, green hydrogen can be sited and scaled more flexibly to serve a variety of purposes. Truck stops could operate their own electrolyzers to produce green hydrogen on site or truck it in from a nearby producer. Wind and solar farms could produce green hydrogen from surplus power and then use it in fuel cells to produce power when needed most, treating hydrogen as a form of on-site energy storage.

Flexible scales and siting of green hydrogen production can limit the need for new pipelines. The United States has just 2,600 kilometers of hydrogen pipelines, mostly serving Gulf Coast refineries and chemical plants, compared to 5 million kilometers of natural gas pipelines. Hydrogen can be blended into natural gas at concentrations of 5 to 15 percent by volume, but moving to higher percentages or pure hydrogen would require repurposing old gas lines or building new hydrogen ones. Even with those pipelines, hydrogen's lower energy density per volume means that more energy would be needed to compress and pump it. That would make long-distance transport of hydrogen uneconomic, but fortunately it is not necessary. "We don't need such long-distance pipelines because the sources of green hydrogen are already naturally distributed, like wind in the Midwest and solar in sunny regions," Brouwer told me. Tanker trucks could be complemented by a local network of hydrogen distribution lines, perhaps repurposed from existing gas systems, to distribute hydrogen locally.[45]

Synthetic Fuels

Despite its many merits, hydrogen has shortcomings. It doesn't pack much energy per volume. It's tough to transport. And it can't be used in most existing engines. However, hydrogen can be transformed into denser synthetic fuels that are more readily transported and used.

Synthetic hydrocarbons could take their hydrogen from electrolysis and their carbon from carbon capture or biomass, making them nearly net-zero fuels. Green hydrogen and the synthetic fuels made from it are collectively known as "electrofuels," or the "X" of "Power to X," since their production is powered by electricity.

Synthetic fuels can be designed for "drop-in" use in engines that would normally burn petroleum-based fuels. For example, synthetic n-octane could replace jet fuel, and other synthetic fuels could replace diesel or marine fuel. Synthetic methanol or ammonia could be used in fuel cells or engines. All these fuels pack much more energy per volume than gaseous hydrogen, enabling them to power heavy modes of transportation.[46]

Rather than using it to synthesize liquid fuels, hydrogen could instead be reacted with carbon dioxide and carbon monoxide to form methane gas in a process known as methanation. That would enable it to be distributed and used in traditional gas systems. However, the inefficiencies of electrolysis and methanation mean that over half of the energy is lost in the conversions. "Methanation is so expensive energy-wise," Webber said. Yet based on his experience at Engie, a utility that provides both gas and electricity, Webber advocates synthetic methane in locations where heating cannot be easily electrified.

"It's going to be a lot cheaper to leverage what we've already invested in natural gas systems rather than replace them with hydrogen," said Webber. "If we go to hydrogen, the pipelines, compressors, storage tanks, appliances—everything is more expensive." Still, given the inefficiency of producing it, the leaks from transporting it, and the pollution from burning it, synthetic methane is a poor substitute for electric heating in all but the coldest regions.[47]

Biofuels

Biomass provides about 5 percent of the U.S. energy supply today, about half of it in the form of liquid biofuels, mainly ethanol and biodiesel. Most of the rest comes from wood and related materials burned for heat and electricity, and biogas captured from landfills, wastewater treatment plants, and dairy farms.[48]

Because the carbon in biofuels was recently captured from the air by photosynthesis, burning it represents a net-zero carbon cycle. However, that does not count greenhouse gases emitted along the way. Adding up the nitrous oxide released from fertilized soils, the gas used to make fertilizers, the diesel used in farming and delivery, and the gas and electricity used to run the biorefinery, many analysts find that traditional ethanol and biodiesel are little cleaner than gasoline or diesel. Destroying forests to create fields for bioenergy crops compounds the problem. Furthermore, powering cars and trucks with biofuels rather than electricity or hydrogen means that their tailpipes release air pollution near where pedestrians breathe it most.[49]

Roughly 40 percent of corn and 30 percent of soybean oil in the United States are already devoted to making ethanol and biodiesel, which are then blended into gasoline and diesel. That leaves little room to scale up, unless new biofuels are made not from food crops but from algae, grasses, or residues from agriculture or forestry. Such materials could triple bioenergy without curtailing demand for food, feed, fiber, or timber, according to DOE. However, given its huge land and water needs and the challenges of gathering materials with low energy density, bioenergy production will ultimately be limited.

Thus, bioenergy should be targeted toward items that are difficult to electrify, such as airplanes and ships.[50]

The federal Renewable Fuel Standard enacted by Congress in 2005 and expanded in 2007 called for quadrupling biofuel use in vehicles by 2022. Most of the growth came from corn ethanol at first, but any growth beyond 2015 was supposed to come from non-food biomass. Congress expected advanced biofuels to be made from feedstocks like cellulose. However, as advanced biofuels have so far failed to scale up, EPA has repeatedly weakened the standard to avoid disruptions in fuel markets. That undermines the "demand pull" that the policy was expected to provide and raises doubts about the future of advanced biofuel technologies.

The quest for advanced biofuels continues. Audi is cultivating microorganisms that use sunlight to produce e-diesel and e-ethanol from carbon dioxide and water. DOE and companies such as Exxon-Mobil are investing in RD&D for producing biofuels from algae, which would avert most of the land, water, and fertilizers needed for crop-based biofuels. None of these can yet match the price or scale of petroleum-based fuels or traditional biofuels. However, prices are getting close enough that a revitalized demand pull could drive a virtuous cycle of improving technologies, performance, and cost.[51]

Biogas can be captured from decomposing waste at landfills, wastewater treatment plants, and dairy farms. Thousands of such facilities already capture biogas, with much of it burned on-site for electricity rather than injected into pipelines. Since biogas is mostly methane, capturing it not only provides energy but also keeps a potent greenhouse gas out of the air.

Even if all waste facilities captured their biogas, that would provide only one percent of the current U.S. gas supply. Cellulosic material from cropland, forests, and specially grown energy crops could theoretically yield ten times as much biogas. However, that would take cellulose away from other biofuel production and require costly technologies. Thus, biogas is unlikely to substitute for more than a sliver of natural gas.[52]

Bioenergy could in theory be paired with carbon capture as a negative emissions technology, since the carbon was recently fixed from the air by photosynthesis. This concept, known as biomass energy with carbon capture and storage (BECCS), has already been demonstrated at ethanol plants in the United States and Canada, but only at small scales. I will consider BECCS alongside other negative emissions technologies in Chapter 7.[53]

America's Role

How could the pursuit of electrification and fuel switching in the United States inform and accelerate parallel pursuits abroad?

The United States is blessed with natural resources that ease our pursuits. Abundant farmland and forests and a relatively low population density provide more biomass per person than in most of the world. An exceptionally large proportion of the country sits atop geology well-suited for carbon storage. Thus, biofuels and blue hydrogen will be easier to produce in the United States than in much of the world. Abundant wind and sunshine can power the production of green hydrogen and synthetic fuels derived from it. In fact, Elizabeth Connelly, who researched hydrogen at NREL, foresees an opportunity for the United States to export hydrogen or synthetic fuels to industrialized countries like Japan, where winds and sunshine are weaker.[54]

The United States lags in some key technologies. China has pushed far ahead in electric cars and buses and the lithium-ion batteries that power them, producing more of each than the rest of the world combined. Europe leads the world in expanding the use of hydrogen.[55]

Where the United States excels is in making clean technologies seem cool, from Tesla cars to Nest thermostats. Furthermore, our innovation ecosystem, from universities to national laboratories to industry to venture capitalists and other funders, is second to none. Dedicating that ecosystem to making electrification more affordable and attractive could catalyze decarbonization beyond our borders. "By far the highest leverage things we can do are develop technologies and demonstrate how you actually decarbonize," Sivaram told me. That will require a massive scale-up in funding for clean energy RD&D, prioritizing projects such as Hydrogen Shot and electric vehicle development that could make decarbonization more affordable and attractive abroad. With emissions growing most rapidly in India, Africa, and Southeast Asia, it will be crucial to develop products that perform well in hot climates and that are affordable to low-income households.[56]

As the world's largest consumer market, the United States also has leverage to drive a push toward decarbonization around the world. American purchases of electric vehicles, heat pumps, and other electrified items could create a demand pull that would influence manufacturing decisions worldwide. Through climate clubs and trade deals, the United States can work with other countries to ensure that low-carbon manufacturing methods are favored rather than disadvantaged in the marketplace, and it can make clean energy technologies available in poorer countries.

The more these approaches can drive down emissions, the less burden will be placed on the remaining pillar of decarbonization, negative emissions technologies.

seven GOING NEGATIVE

However deeply we cut emissions, they won't reach zero. Airplanes, ships, and some industries will be difficult to decarbonize. Not all buildings or equipment will be retrofitted or replaced. Some greenhouse gases like nitrous oxide don't have engineered sinks. Thus, getting to net-zero greenhouse gases—as we must to stabilize the climate, and as the Paris Agreement committed the world to do within this century—will require new sinks to offset the emissions that remain. In other words, we will need to remove carbon dioxide from the atmosphere via negative emissions technologies or other means. That raises thorny questions about which approaches to pursue and the impacts of using them.

Decarbonizing energy and averting emissions are often a win-win situation. Energy efficiency and clean energy reduce not just climate-warming emissions but also the water and air pollution and ecosystem degradation that come from extracting, refining, transporting, and burning fossil fuels. Efficiency saves more money than it costs. Clean electricity and electrification can save money too, as renewable electricity, electric cars and buses, and heat pumps become cheaper than their rivals. Preventing leaks of methane and accompanying light hydrocarbons improves air quality. Averting nitrous oxide and CFC emissions protects the ozone layer.

By contrast, apart from forestry and agriculture, negative emissions technologies can be expensive or environmentally damaging, yielding little benefit beyond the carbon removal itself. Most would require grinding, burning, processing, and transporting massive amounts of materials, using vast resources of energy, land, and water to do so. Such capital-intensive activities may create few jobs and damage the environment locally for the sake of the global climate, raising concerns for environmental justice and local opposition.

Consequently, hopes of soaking up emissions later should not dampen the urgency of cutting them now. But there is a real risk that we will do just that. It is easier to wish for future silver bullets than to take action today. If negative emission technologies fail to scale up and emissions remain unabated, warming will continue apace.[1]

Nevertheless, carbon dioxide removal will be essential to stabilizing the climate. Net zero requires sinks to balance sources. Most scenarios for holding warming inside the Paris Agreement limits rely on massive removal of carbon dioxide. How much we'll need depends on how much we emit and the still uncertain sensitivity of climate to those emissions. For a 1.5°C warming limit, the "middle-of-the-road" pathway highlighted by IPCC's influential 2018 special report relied on 50 gigatons (1 gigaton = 1 billion metric tons) of carbon dioxide removal cumulatively by 2050 and a bit over 500 gigatons by 2100. For a sense of scale, 50 gigatons is about as much carbon dioxide as is emitted by fossil fuel burning worldwide every eighteen months, and 500 gigatons is enough to warm the world by about 0.25°C. For 2°C, IPCC scenarios included similar amounts of carbon removal, with the laxer temperature target allowing emissions to remain higher. This means the global needs for carbon dioxide removal in most scenarios start small and scale up to roughly 10 gigatons per year later this century.[2]

Apportioning those targets by nation is tricky. The Paris Agreement states that net-zero greenhouse gases globally should be reached with "equity and common but differentiated responsibilities and respective capabilities, in the light of different national circumstances." As the source of 13 percent of global emissions today and 25 percent of cumulative emissions historically, the United States would be "responsible" for around 20 percent of carbon dioxide removal. "National circumstances" make the United States capable of shouldering an even larger share, given the country's technological and financial prowess and its vast forests and farmland and geological formations where carbon could be stored. Therefore, scaling up carbon dioxide removal to around 2 gigatons per year, as the U.S. share of a 10 gigaton per year global target, is within the realm of what may be needed for net zero.[3]

Natural sinks pull about 12 gigatons of carbon dioxide into vegetation and soils, 9 gigatons into oceans, and 1 gigaton into rock weathering each year. That doesn't count toward the new sinks we need, but it provides a sense of scale. The vast land and ocean sinks have long in-spired ideas for expanding them through agriculture, forestry, ocean fertilization, or other means. More recent technologies aim to radi-cally accelerate rock weathering or suck carbon dioxide from thin air. This chapter explores these options before turning to a more radical option—geoengineering incoming sunlight to reset the planet's ther-mostat.[4]

Forestry and Agriculture

American forests already soak up 0.8 gigatons of carbon dioxide each year, mostly from the growth of trees in existing forests. Rising levels of carbon dioxide—the airborne "food" that plants "eat" via photo-synthesis—are accelerating that growth. American forests are also ex-panding across the map, reclaiming former cropland and grassland faster than they are destroyed by swelling cities.[5]

Forestry could accelerate carbon uptake to provide some of the new sinks that are needed. A panel of leading experts convened by the Na-tional Academies of Sciences, Engineering, and Medicine estimated that it would be feasible to sequester an additional 0.15 gigatons of carbon dioxide per year by planting more trees in the United States (Table 1). However, doing so would require converting lands the size of New Jersey and Connecticut combined into forests. An additional 0.1 gigatons per year could be sequestered in existing U.S. forests by improving forest management practices, such as protecting trees from insects and disease and waiting longer to harvest timber. That could improve local habitats beyond the benefits to global climate. How-ever, carbon sequestered by forests is also at risk of going up in smoke as wildfires intensify.[6]

Soils contain far more carbon than the atmosphere and vegetation combined. Although some carbon stays in the soil for centuries, car-bon near the surface can return to the atmosphere within only months or years. Like forests, soils may provide only temporary storage of carbon.

Agriculture can help soils store more carbon and retain it longer. Cover crops such as grasses or legumes, planted during the off-season

Table 1. Safe potential rate of carbon dioxide removal (in gigatons/year) in the United States with current technology at costs under $100/ton

Negative Emissions Technology	Potential (Gt/year)
Afforestation and reforestation	0.15
Forest management	0.10
Agricultural soils	0.25
Bioenergy with carbon capture and storage (including biochar)	0.50
Weathering	Unknown
Direct air capture	0.00
Total of quantified options	1.00

Source: National Academies of Sciences, Engineering, and Medicine, *Negative Emissions Technologies and Reliable Sequestration*, Table S.1

when fields would normally be bare, can photosynthesize carbon dioxide from the air while protecting the soil from erosion. Less than 5 percent of American farmers plant cover crops today, since they may not yield much added profit relative to the extra work involved. So there is plenty of room to expand cover crops, but incentives may be needed. Farmers could also rotate longer-lasting crops such as hay with shorter-lived crops such as corn to accelerate carbon uptake. Reducing how often fields are tilled can help carbon stay in the soil longer. Taken together, these farming practices could boost carbon dioxide uptake by about 0.25 gigatons per year, according to the National Academies panel. Better management of grazing lands could sequester additional carbon, but the panel deemed this too uncertain to quantify.[7]

Another option for farmers is to apply biochar to their soils. Biochar is a charcoal-like substance that can be made by charring crop residues in the absence of oxygen, a process known as pyrolysis. That can lock up carbon far longer than if the residues were merely left to decompose. Biochar can provide a triple win by sequestering carbon, boosting crop yields, and reducing the need for fertilizers. Fertilizers are made mostly from natural gas and are the leading source of nitrate runoff to streams and ammonia and nitrous oxide emissions to the air.

Reducing their use would therefore benefit the water, air, and climate. The longevity of carbon sequestered by biochar remains uncertain, but much of it can remain in the soils for centuries.[8]

Scaling up any of these agricultural measures will not be easy, since they depend on actions by hundreds of thousands of farmers. Gains in crop yields may be too small to motivate farmers to change their practices, so farmers may need to be paid for the carbon capture. That will require better methods to measure carbon capture and its longevity in soils.

Bioenergy with Carbon Capture and Storage

Bioenergy with carbon capture and storage (BECCS) involves producing electricity or other energy from biomass while sequestering the carbon. Because carbon in biomass was recently photosynthesized from the air, capturing it would provide a net sink. Climate modelers in the early 2000s began relying on BECCS as the dominant engineered sink or "savior technology" in their scenarios, imagining that it would somehow be scaled up to balance or even exceed remaining emissions. "In little more than a decade, BECCS had gone from being a highly theoretical proposal for Sweden's paper mills to earn carbon credits to being a key negative emissions technology underpinning the modeling, promoted by the IPCC, showing how the world could avoid dangerous climate change this century," wrote Leo Hickman in chronicling the curious rise of BECCS to savior status. Lacking more realistic means for achieving temperature targets amid rising emissions, climate modelers have more recently added other untested but less land-intensive sinks such as direct air capture as placeholders for imagined carbon dioxide removal.[9]

Despite its imagined savior role, BECCS so far operates only at a handful of ethanol plants, capturing modest amounts of carbon dioxide. A few other deployments are planned at biomass and waste-to-energy power plants. Even if those pilot projects prove viable, any scale up of BECCS will ultimately be limited by the amount of biomass that can be gathered affordably without constricting land needed for food, fiber, timber, and wildlife. The National Academies panel estimated that 0.5 gigatons of carbon dioxide per year could be se-questered via byproducts from U.S. agriculture and forestry, such as corn stover and sawmill waste. Cultivating bioenergy crops such as switchgrass or miscanthus could more than double that potential, but with far more use of land, irrigation, fertilizers, and farm equipment. Burning the biomass for electricity while capturing the carbon could provide a flexible power source to balance variable wind and solar. However, since most forests and farms lie atop geology that is ill-suited for underground storage, pipelines would be needed to pump carbon dioxide to repositories such as saline aquifers or depleted oil reservoirs.[10]

The appeal of BECCS power plants has diminished as renewable electricity has grown cheaper, although such facilities could garner premium prices by filling in gaps between demand and variable wind and solar supply. "Power plant BECCS doesn't make sense now that renewable electricity is so cheap, because it's no longer a strategic use of limited biomass resources," said Jim Williams, who models deep decarbonization scenarios. Biomass may be more valuable as a feed-stock for fuels for aviation or shipping, which are difficult to electrify. Capturing carbon dioxide from biofuel manufacturing processes as is done at the ethanol plants with BECCS can reduce the carbon foot-print of biofuels.[11]

Strengthening the Ocean Sink

As seawater warms, its ability to hold carbon dioxide diminishes, like a warm Coke losing its carbonated fizz. Although seawater is naturally alkaline thanks to dissolved minerals, absorbing carbon dioxide re-duces that alkalinity, acidifying seawater and impairing its ability to absorb carbon dioxide. Warming and acidification also imperil shell-fish and corals, which can survive only narrow ranges of temperature and acidity.

Restoring the alkalinity of surface seawater would rejuvenate its up-take of carbon dioxide from the air. Scientists have proposed spraying materials such as lime onto oceans to boost alkalinity. However, kilns need 900°C heat to make lime from limestone, an energy-intensive process that pushes overall sequestration costs to around $100 per ton. Grinding mine tailings, steel slag, or silicate rocks may offer cheaper and less energy-intensive options. Calcium and magnesium in those materials would lock up carbon in compounds like calcium carbonate, the building blocks of shells and coral. Meanwhile, their silica and iron could fertilize photosynthesis by phytoplankton. Together, that could sequester about two tons of carbon dioxide per ton of silicate rock that is pulverized and sprayed on seawater. A gigaton per year sink would thus require grinding up around 500 million tons of rocks, nearly the scale of U.S. coal mining today, plus hundreds of ships to spread the powder at sea.[12]

Whereas these alkalinity-based approaches deserve further research, fertilizing the ocean by instead applying iron, nitrogen, or other nutrients is more problematic. Such fertilization aims to stimulate photosynthesis by phytoplankton. When phytoplankton and the animals that eat them die and fall to the ocean floor, they take the carbon with them. However, when they decompose closer to the surface, the carbon can quickly return to the air. More worrisome, algal blooms from excess fertilization can release toxins and consume oxygen as they decay, devastating nearby ecosystems.

A critical review of carbon sink strategies concluded that ocean fertilization with iron or nitrogen is "not a viable" option due to sustainability concerns. Other scientists have argued that earlier tests of iron fertilization were so damaging that even pilot studies should be avoided. Concerns about ocean fertilization have prompted moves to restrict testing of all carbon removal strategies at sea under the London Protocol on Marine Pollution. That could stymie field testing of the alkalinity-based approaches that, unlike iron or nitrogen fertilization, counteract ocean acidification, potentially providing a win-win for marine ecosystems and climate.[13]

Engineered Weathering on Land

Rather than spraying mineral-rich powder on the oceans, some scientists have suggested spreading it across cropland. That would counteract soil acidification and supply nutrients to crops, thereby boosting crop yields while reducing nitrogen fertilizer use and all the damage that comes with it. As the calcium and magnesium react with carbon dioxide from the air, they would bind it up as carbonates that would remain in the soil or flow to the oceans. Scientists estimate that applying this approach across 10 percent of U.S. croplands could remove 0.1 gigatons of carbon dioxide per year at a cost of $160 per ton, before accounting for the benefits to farmers. That's a hefty price, and it's probably a stretch to imagine that so many farmers would adopt an unfamiliar practice unless incentives or other benefits are substantial.[14]

Another option is to spread a magnesium oxide powder over barren ground rather than cropland, wait for it to react with carbon dioxide from the air, and then heat the resulting compound to regenerate the magnesium oxide and capture the carbon dioxide. That would allow for many cycles of carbon capture from each ton of rock pulverized, at a cost of around $50 per ton. Sequestering one gigaton per year by this method would require roughly 7,500 square kilometers, most of it covered by 10 centimeters or so of the powder and the rest by solar panels to power the heating. That is an enormous amount of land, but just half the size of the Nevada Test and Training Range. "All of this can be done with existing technologies at current prices," said Peter Kelemen, a geochemist at the Lamont-Doherty Earth Observatory who helped develop the concept.[15]

With so many strategies theorized but none demonstrated at scale, and with risks and benefits poorly understood, the prospects for engineered weathering remain unclear. An influential meta-analysis in 2018 estimated a global potential of 2–4 gigatons of carbon dioxide per year, at costs of $50–$200 per ton. The National Academies panel in 2019 declined to quantify opportunities in the United States but concluded that they "could have very large capacity if their costs and environmental impacts could be sufficiently reduced"—big ifs for novel techniques. The panel proposed an ambitious agenda to boost scientific understanding and fund pilot projects. "We know these reactions are occurring spontaneously in nature on geological time-scales and removing carbon dioxide for free, but what we don't know yet is if our ideas will work to scale this up on human time scales," Kelemen told me.[16]

Direct Air Carbon Capture and Storage

Direct air carbon capture and storage (DACCS) aims to suck carbon dioxide out of thin air. That's akin to separating a needle from a haystack or a few drops of food coloring from a pitcher of water, since air contains just one carbon dioxide molecule for every 2,400 molecules of everything else. To begin the process, fans push air across a solid sorbent or a liquid solvent that locks up the carbon. Heat is applied to release the carbon dioxide, which must then be compressed, transported, and stored. All of this requires electricity and heat, negating much of the carbon capture unless the electricity and heat are produced cleanly. Unfortunately, some proposals for DACCS would use natural gas or even coal with carbon capture to supply the energy, increasing the need for fracking or mining and transport of fossil fuels. Even with cleaner energy sources, it's unclear whether DACCS is a wise use of that energy and associated resources. Only if electricity and heat become exceptionally clean, cheap, and abundant will DACCS be worth considering.[17]

As doubts have emerged about the viability of BECCS and its impacts on ecosystems and cropland, DACCS has supplanted it as a presumed savior technology in models and the public imagination. In practice, though, pilot projects have so far been tiny and expensive, and their benefits are not clear-cut. Climeworks opened the world's first commercial direct air capture plant in Switzerland in 2017, relying on power and heat from an adjacent waste incinerator. Its iconic array of fans captures just a few hundred tons of carbon dioxide per year for use by a nearby greenhouse. Another Climeworks facility in Switzerland converts carbon dioxide into methane to be burned. Only its facility in Iceland yields a durable sink for carbon dioxide, converting it into rock. Costs for carbon dioxide capture by Climeworks plants have been reported at $600 per ton, and the company charges $1,100 per ton to subscribers who crowd-fund its efforts.[18]

Another company, Carbon Engineering, is building its first pilot facility in Squamish, Canada. It expects costs comparable to those of the Climeworks plants at first. However, Carbon Engineering officials estimate that their liquid solvent-based design could eventually scale up to capture carbon dioxide for $113 per ton in the configuration that the National Academies panel deemed most plausible. Each million-ton-per-year facility would use as much gas as 80,000 gas-heated homes, plus the electricity output of a 500-hectare solar farm (a hectare is about the size of two-and-a-half football fields). We would need 1,000 such plants to capture 1 gigaton per year, gobbling up 10 percent of current U.S. energy consumption without serving a useful purpose beyond the carbon capture itself.[19]

Summing Up the Sinks

With dozens of approaches theorized and none yet deployed at scale, it is hard to predict which options will prove most effective or what the ultimate costs of carbon removal will be. Some approaches that seem promising in pilot projects will ultimately fizzle, while others that seem far-fetched today may prove viable with clever tweaks and innovations. Predicting winners today is as wise as predicting Nobel laureates from today's graduate students. Nevertheless, some broader points should be noted.

First, win-win opportunities should be pursued post-haste. Planting trees on contaminated brownfields, sowing cover crops on eroding farmland, and adding carefully designed biochar to nutrient-deficient soils would enhance ecosystems, slow erosion, and boost crop yields while cutting fertilizer use and soaking up carbon. In all these cases, ancillary benefits could outweigh costs, even if carbon sequestration goes unrewarded. Climate benefits are icing on the cake.

Unfortunately, many potential carbon sinks could do more harm than good beyond their climate impacts. Enhanced weathering would require enormous amounts of mining and grinding and could spread toxic metals across soils or oceans. BECCS may be benign if the biomass is grown on barren or marginal lands but problematic if homogenous fields of energy crops replace farmland or diverse ecosystems. BECCS would also require machinery to move biomass to power plants plus pipelines and repositories for the captured carbon dioxide. DACCS would consume enormous amounts of energy and other resources while yielding little benefit beyond the carbon capture itself.

At $100 per ton, in line with the costs targeted by some negative emissions technologies, a U.S. goal of 2 gigatons per year would cost $200 billion per year. That is about 1 percent of U.S. gross domestic product, or around 4 percent of the federal budget—not unfathomable, but still an onerous expense. By comparison, President Biden's budget proposal for 2022 sought to scale up federal climate-related investments to $36 billion, including $10 billion for nondefense clean energy innovation. Unlike carbon sinks, efficiency and clean energy can pay for themselves over time and yield benefits beyond climate, warranting immediate investment. Negative emissions technologies belong more in the realm of R&D for now, to expand options and bring down their costs ahead of widespread deployment.[20]

Pilot deployment of carbon sequestration in the United States is being driven by tax credits established under Section 45Q of the Internal Revenue Code. Such credits create a "demand pull" to stimulate innovations but can also result in perverse outcomes. For example, plans to replace a coal power plant in New Mexico with solar were put on hold while the plant's owner sought 45Q tax credits plus a DOE grant to install carbon capture instead. Solar power is far cleaner and cheaper than keeping the coal plant running, but it does not qualify for 45Q tax credits or a DOE grant. Policies should be redesigned to not favor capturing emissions over avoiding them in the first place.[21]

Voluntary efforts can create their own demand pull. Hundreds of companies, ranging from Amazon to JetBlue to BP and Shell, have set net-zero mandates for their operations. Microsoft has gone even further, pledging to offset prior emissions with new sinks. Numerous cities, universities, and other entities have set net-zero targets too. Sinks should be verified to ensure that net reductions are achieved without harming local communities or ecosystems.

Environmental justice must not be neglected. Some carbon sinks require intense activities such as chopping and burning biomass, grinding rocks, sucking carbon from the air, and pumping it to underground repositories. Too often, activities like this that impose local burden for the greater good have been sited in vulnerable communities, with too little done to mitigate local harms. Landfills, metal recyclers, and wastewater treatment plants fit that unjust pattern. Stringent regulations will be needed to ensure that carbon capture is conducted away from sensitive communities, with local harms minimized.

Above all, hopes for negative emissions technologies in the future should not forestall actions to mitigate emissions now, a moral hazard known as "mitigation deterrence." Substituting emissions removal for emissions mitigation would mean missing out on the health, environmental, energy security, and other benefits that accompany shifting away from fossil fuels. If mitigation is deterred and alternative interventions don't materialize, society could be locked in to dangerous amounts of warming.[22]

Solar Geoengineering

The moral hazard of mitigation deterrence—of inaction in hopes of a future silver bullet—looms even larger for solar geoengineering. Nevertheless, the daunting scales needed for carbon sinks and the failure of mankind to rein in emissions have left some scientists searching for shortcuts to slow warming. The option considered most is solar geoengineering—reflecting sunlight away from Earth.

Solar geoengineering would reset the thermostat on a warming planet. Accumulating greenhouse gases are trapping in ever more of Earth's warmth. Injecting particles into the stratosphere would counteract some of that greenhouse warming by reflecting away some sunlight. Injection into the stratosphere lets particles remain above the highest rain clouds for a year or two before settling out, similar to what occurs after major volcanic eruptions. The last such eruption, of Mount Pinatubo in the Philippines in 1991, raised stratospheric particle levels enough to cool temperatures by more than half a degree Celsius for a little over a year.[23]

Scientists estimate that just a few billion dollars per year could be enough for specially designed high-altitude tanker planes to inject enough material into the stratosphere to offset half of new warming. That price tag is trivial compared to the trillions of dollars likely to be needed globally to slash fossil fuel use and capture the emissions that remain. That makes solar geoengineering an appealing shortcut to reduce the need for costlier emissions mitigation or removal.[24]

However, extreme caution is needed before we consider this get-out-of-jail-free card. Solar geoengineering might be able to offset global warming, but it would create global weirding instead. Stabilizing the thermostat would require injecting ever more particles into the stratosphere each year, because particles settle out annually while carbon dioxide accumulates for centuries. Any pause in the injections would immediately jolt the climate warmer, rocking societies that have adjusted to the climate's geoengineered norms.

On the other hand, if stratospheric injections of particles continue to grow to keep pace with ever more carbon dioxide below, more and more sunlight would be reflected off the particles into space while more and more of Earth's infrared radiation would be held in the troposphere. That might hold the global thermostat steady, but it would warp everything else about our climate. Skies would be dimmer, albeit with more brilliant sunsets. Output from solar panels would fade with diminished sunshine. Concentrated solar power would become impractical, as less sunlight would arrive in the direct beam reflected by mirrors. Evaporation depends not only on temperature but also sunlight, so it would slow. What goes up must come down, so rainfall would slow too. Some places would get wetter and others drier, as jet streams shift in unpredictable ways. Those shifts along with changes in ocean circulation would make some areas cooler and others warmer, even if geoengineering stabilizes temperatures on a global average basis. Oceans would continue to acidify as carbon dioxide levels climb.

Some climate models predict that solar geoengineering would actually moderate climate hazards, because slower evaporation would alleviate droughts and diminished rainfall would alleviate flooding and lessen the severity of tropical cyclones. Maybe so. But modern society has been built around the climate of recent decades. No matter how numerous the winners, the losers from global weirding will have strong grounds for anger or retaliation. It's impossible to predict who the losers will be and how badly they will be hurt—whether by diminished rainfall or damaged habitats or anything else—and how they will respond to the damage. Behavioral economics teaches that losses are experienced more acutely than gains. Thus, fervent opposition is sure to arise from those who perceive themselves to be harmed.[25]

All of these complications make the low cost of solar geoengineering as much a curse as a blessing. A few billion dollars a year to operate a fleet of stratospheric planes is well within the means of dozens of countries and even some multi-billionaires. Who is to say who gets to reset the thermostat and transform the weather for the entire planet? Perhaps heat-stricken countries or island nations would want to make Earth cooler than today, while colder countries might welcome more warming. Some countries would get wetter and others drier. The world has not begun to fathom how to govern such capabilities.

Even if solar geoengineering is never deployed, the sheer prospect of it on the horizon raises serious concerns about mitigation deterrence. Will we act to cut emissions and create carbon sinks as urgently if we imagine a shortcut to a cooler planet? Will we invest trillions of dollars in clean energy and carbon sinks, when mere billions could throttle the thermostat?[26]

No mirage of a panacea should distract us from slashing emissions as quickly and deeply as possible and offsetting what's left with new sinks. What policies are needed to make that happen?

eight CONFRONTING POLICY GRIDLOCK

Breakthroughs in diplomacy and technology have brought us closer than ever to confronting climate gridlock. But policy has lagged. Without bold policy, the opportunities opened by diplomacy and technology will go unseized. A net-zero future and clean energy economy will fall further out of reach.

The Paris Agreement together with emerging ideas for climate clubs can leverage each country's actions to jump-start progress abroad. But there will be nothing to leverage if a country does not act. Without policy, diplomacy is not credible. Once a country does act, embedding its policies in its Paris commitment and climate club obligations can make them more durable and encourage others to act too. With its market size, technological prowess, and clout, the United States has an unparalleled ability to drive the creation of climate clubs and spur other nations to do more.

Clean energy and negative emissions technologies are undergoing rapid innovations, but they are not being deployed at anywhere near the rates needed to reach net zero. Fossil fuels continue to power most of the economy in the United States and globally, much as they have for the past century. Without new policies, energy forecasters project that fossil fuels will continue to dominate, and emissions will plateau for decades to come.[1]

Optimists may hope clean technologies will mimic the most abrupt transitions of the past. Cars replaced horses, color televisions replaced black-and-white ones, and smart phones replaced flip phones within a decade or two, driven more by market forces than policy. Each tracked an S-shaped adoption curve, with slow early adoption followed by a leap to dominance and then ultimately market saturation.[2]

Unfortunately, most clean energy and negative emissions technologies will not attract such rapid adoption via consumer appeal alone; policy will be imperative. More efficient lightbulbs, appliances, vehicles, and buildings all provide lifetime savings, but cost more upfront.

Technologies for avoiding methane leaks, replacing HFCs, and mitigating nitrous oxide are all readily available, but without policy there is little self-interest in deploying them. Wind and solar are the cheapest new sources of electricity, but they must compete with legacy coal and gas plants whose capital costs are paid. Electric vehicles are reaching cost parity with fossil-fueled ones, but they struggle with customer and dealer wariness and lack of charging infrastructure. Electrification of heating and industry proceeds far too slowly in the absence of adequate policies. Green hydrogen could provide a versatile feedstock and fuel, but for now it suffers from high costs and lack of distribution infrastructure. Negative emissions technologies remain in their infancy and are unlikely to be deployed without incentives or mandates. Across these and other technologies needed to mitigate climate change, the International Energy Agency has concluded that the vast majority are emerging far too slowly to achieve Paris Agreement targets.[3]

Policies can push and pull technologies toward faster diffusion. The "technology push" comes from research and development that drives innovations, and the "market pull" comes from policies that boost demand for those technologies. Push and pull policies are inherently synergistic, driving technologies along their learning curves toward better performance and lower costs with broader adoption.[4]

Diplomacy, technology, and policy together can create a virtuous cycle. The Paris Agreement and climate clubs motivate domestic policies to meet the expectations of other countries and leverage domestic advances beyond them. Those policies drive a push for technology R&D and a market pull to adopt those technologies, reducing their costs. Technological gains in turn make it feasible to pursue more vigorous policies and diplomacy. Although inadequate policies historically have failed to complete such a cycle, emerging conditions open promising pathways ahead.

History of U.S. Climate Policy

Faced with the grandest environmental challenge of our times, the U.S. Congress has repeatedly failed to enact comprehensive legislation confronting climate change. Congress has instead passed piecemeal measures promoting energy efficiency, funding RD&D, and providing tax credits for renewable energy, electric cars, and carbon capture. EPA has shoehorned climate-oriented regulations into landmark acts that Congress passed decades ago for other purposes, before global warming became a top concern. Those acts include the National Environmental Policy Act of 1969, which did not incorporate climate impacts into environmental assessments until the 1990s; the 1972 Clean Water Act, which considers climate change only tangentially and neglects the effects of carbon dioxide on ocean acidification; and, most importantly, the Clean Air Act.[5]

Passed in 1963 and extended with major amendments in 1970 and 1990, the Clean Air Act provides the foundation for air pollution policy in the United States. The act was traditionally assumed to apply only to conventional air pollutants such as sulfur dioxide, carbon monoxide, particulate matter, and airborne toxics that directly endanger public health and welfare locally. Greenhouse gases, which warm climate globally but are not toxic to breathe, were mostly neglected. Through regulations on vehicles and industry, standards for ambient air quality, cap-and-trade programs for power plants, and other measures, the Clean Air Act helped drive down air pollutant emissions by 77 percent from 1970 to 2019, even as carbon dioxide emissions rose.[6]

Regulation of greenhouse gases through the Clean Air Act took root in 2007, when the Supreme Court ruled 5–4 in *Massachusetts v. EPA* that the act obligates EPA to regulate greenhouse gases if it determines that they endanger public health and welfare. EPA soon determined that they do. That finding paved the way for EPA to impose carbon dioxide emissions standards on vehicles and to regulate methane leaks.[7]

Despite that ruling, EPA has struggled to shoehorn greenhouse gas regulations into the decades-old Clean Air Act. Under President Obama, EPA issued the Clean Power Plan, setting state-by-state limits on power plant emissions. However, the regulations never took effect due to litigation, and the limits were so weak that they were achieved without the plan itself. The Trump administration repealed and replaced the Clean Power Plan with a toothless rule, which itself was blocked by litigation. Rules on vehicle emissions and methane leaks have been similarly plagued by litigation and reversals. A ban on some

HFC refrigerants was overturned in 2017 by soon-to-be Supreme Court justice Brett Kavanaugh, who acknowledged the merits of banning these potent greenhouse gases but objected to how the rule was squeezed in to the Clean Air Act. Restrictions on those refrigerants arose instead from new legislation passed by Congress in 2020.[8]

In sum, shoehorning regulations into decades-old legislation has proven to be inadequate both for combating climate change and for giving businesses and consumers the clarity they need to make prudent decisions. Yet Congress has for decades failed to pass comprehensive climate legislation, despite coming close on two occasions.

The first near miss came in 1993, after newly elected President Bill Clinton implored Congress to enact a tax on energy. Technically, this was not a climate proposal, as Clinton never even uttered the word "climate" in his address. Instead, he pitched the tax as a way to reduce the budget deficit, pollution, and dependence on foreign oil. The initial tax would have been small, but it could have been escalated over time and set a precedent for taxing fossil fuels. "Had that passed, it would have put the United States in a very strong position internationally, and it would have put the right incentives in place for emission reductions with the most efficient mechanism you can have," recalls Rafe Pomerance, who that year became Clinton's deputy assistant secretary of state for environment and development. Known as the "Btu tax" for taxing the British thermal unit energy content of fuels, the bill squeaked through the House but died in the Senate without receiving a vote. The National Association of Manufacturers spearheaded an anti-tax alliance that grew to be the largest coalition of businesses ever to oppose a single bill. The debacle contributed to the Republican takeover of the House in midterm elections the following year, thwarting any hopes for passing major climate legislation for the remainder of the Clinton presidency. "Getting Btu'd" became synonymous with getting voted out of office after a futile vote.[9]

Opportunity seemed to strike a second time in 2009, when a popular young Democrat once again replaced a President Bush in the White House. Over the previous few years, leaders of environmental organizations, utilities, and other groups had been meeting behind the scenes to build consensus for a cap-and-trade system to limit carbon emissions, hoping that it would win broader appeal than an energy tax. During the 2008 presidential campaign, both the Democratic and Republican nominees, Senators Barack Obama and John McCain, advocated for caps on carbon dioxide emissions. Appearing beside House Speaker Nancy Pelosi in front of the Capitol dome for a television ad, former speaker Newt Gingrich said, "We do agree our country must take action to address climate change." In his first presidential address, President Obama implored Congress to send him "legislation that places a market-based cap on carbon pollution" in order to "save our planet from the ravages of climate change." Congressmen Henry Waxman (D-California) and Edward Markey (D-Massachusetts) obliged, introducing a bill to slash U.S. emissions 83 percent by 2050. Beyond its cap-and-trade provisions, the bill would have provided hundreds of billions of dollars for clean energy technologies, boosted efficiency standards, and subsidized low-income families and energy sector workers.[10]

The Waxman-Markey bill suffered the same fate as Clinton's energy tax, squeaking through the House before dying without a vote in the Senate. Despite the years of behind-the-scenes coalition building by environmental and industry elites, little was done on the left to rally activist support for the bill. "We didn't have strong grassroots, vocal support to counterbalance the artificial megaphones and astroturf that the Koch brothers and others were creating to oppose Waxman-Markey," recalls Eric Pooley, who chronicled the period in his book *The Climate War*. Republican congressional support for what had been envisioned as bipartisan legislation withered in the face of the recession, fears of Tea Party-backed primary challengers, and blistering attacks from the fossil fuel industry and ideological groups. "The climate denial machine kicked into a higher gear than ever, with a heady brew of fear-mongering," said Edward Maibach, who directs George Mason University's Center for Climate Change Communication. Public opinion shifted too, as acceptance of climate science became more partisan. "There was a dramatic collapse of belief in climate change—an exceptional change in opinion," Maibach told me. Democrats again suffered a wave of defeats in the midterm elections, and hopes for major federal climate legislation entered another period of hibernation.[11]

Why Policy Breakthroughs Are Achievable

Despite those historical failures, emerging conditions are opening unprecedented opportunities for policy breakthroughs. The international landscape reshaped by the Paris Agreement provides fertile ground for leveraging domestic advances into stronger commitments abroad. Technologies are making it more feasible and affordable than ever to transition away from fossil fuels.

Trends in public opinion are favorable too. Surveys show that record percentages of Americans accept that global warming is happening and that humans are causing it. Global warming has moved from the realm of polar bears in the future to devastating hurricanes, floods, and wildfires in the here and now. "Public opinion has grown more stable now that people see the impacts in their own communities," Maibach said. Overwhelming majorities of Americans support renewable energy, planting trees, and cutting fossil fuel use. Oddly, even more Americans support restricting carbon dioxide than accept that it warms the climate, suggesting legislation need not wait for education or acceptance of climate science.[12]

However, public opinion is not sufficient to pass legislation. "Let's not kid ourselves that what the average American thinks is what drives policy," said Anthony Leiserowitz, who directs the Yale Program on Climate Change Communication. Research by Matto Mildenberger and Leah Stokes, political scientists at the University of California, Santa Barbara, revealed that Republican politicians and their staffs tend to underestimate their constituents' support for climate action. Those misperceptions are no accident. "Threatened interest groups spend an enormous amount of time and energy shaping perceptions of public opinion," Mildenberger told me. "Many of the social movements and climate advocacy groups are working in opposition to that."[13]

Long-established climate advocacy groups such as the Environmental Defense Fund, Sierra Club, and the Natural Resources Defense Council are now joined by upstarts across the political spectrum. On the left, the youth-led Sunrise Movement rallied support for a Green New Deal, rising to prominence with a sit-in in Rep. Nancy Pelosi's (D-California) office soon after the 2018 midterm elections.

Their website vows to "stop climate change by transforming our whole economy," casting that fight "alongside the fight against white supremacy and colonialism." Rather than sponsoring sit-ins, the centrist Citizens' Climate Lobby emphasizes its "consistently respectful, non-partisan approach," with more than four hundred local chapters advocating for carbon fees and dividends. "The way democracy should work is as a partnership between citizens and elected officials," the group's executive director, Mark Reynolds, told me. On the right, RepublicEn casts itself as a "balance to the Environmental Left" while also advocating for a fee-and-dividend approach. Its executive director, former congressman Bob Inglis, told me that he hopes 15 Republicans in the Senate and 25 in the House can be persuaded to support carbon dividends.[14]

After corporate-funded opposition helped sink the Btu tax and the Waxman-Markey bill, corporations including ExxonMobil have withdrawn support from groups such as the Heartland Institute, which issues climate denialist messaging, and the American Legislative Exchange Council, which crafts anti-environmental legislation. In 2017, big oil and gas companies such as Shell and BP, utility companies such as Exelon and Calpine, and automakers Ford and GM joined major environmental organizations as founding members of the Climate Leadership Council. The Council calls for carbon fees and dividends, gradually rising from a rate of $40 per ton, under a plan written by former secretaries of state James Baker and George Shultz.[15]

In addition to advocating for climate policy, businesses are increasingly "walking the walk" on climate action. Whereas 100 percent clean electricity once set the bar for green pledges, a growing number of businesses are pledging to achieve net-zero carbon emissions not just from electricity but across their operations. As of July 2020, the UNFCCC-backed Race to Zero campaign tallied 995 businesses along with over a thousand other entities that had pledged to go net-zero by 2050. It is remarkable that reaching net zero by 2050 has now become a minimal expectation for corporate and political pledges; in 2016, President Obama's blueprint for an 80 percent cut in net greenhouse gas emissions by 2050 was called "breathtaking" and "hugely ambitious."[16]

Whether these changes in corporate stances pave the way to broader policy remains to be seen. "I don't think large corporations determine what happens in congressional politics," said Theda Skocpol, a sociologist at Harvard University who wrote an influential post-mortem on Waxman-Markey. In the past, public proclamations by corporations have been undermined by behind-the-scenes lobbying by their executives or trade associations. "Unless and until the Chamber of Commerce, National Association of Manufacturers, and Business Roundtable change their stances, Republicans are skeptical about what they hear from oil companies," Jerry Taylor, president of the Niskanen Center, told me in May 2019. Since then, the U.S. Chamber of Commerce has swung around to supporting the Paris Agreement and announced that it "supports a market-based approach to accelerate greenhouse gas emissions reductions across the U.S. economy"; the National Association of Manufacturers called for Congress to act on climate; and the Business Roundtable and American Petroleum Institute announced their support for carbon pricing. Greg Bertelsen, executive vice president of the Climate Leadership Council, told me that their corporate members are matching their public stances with political engagement. "I've personally been in the room with energy executives on Capitol Hill directly pushing for carbon pricing and climate legislation," he told me.[17]

Beyond the lobbying of traditional energy giants, renewable energy producers and their trade associations could gain clout as their businesses grow. Historically, these entities have failed to muster anything close to the political influence of electric and gas utilities. "Utilities and their trade associations put a lot of money into lobbying and build up long-term relationships with legislators and regulators," said Stokes, whose book *Short Circuiting Policy* chronicles battles over energy policy at the state level. Renewable energy trade associations are "fledgling and underfunded," she told me in 2018, and they are fragmented across different types of energy such as wind, utility-scale solar, and rooftop solar. Danny Cullenward and David Victor, in their 2020 book *Making Climate Policy Work*, dismissed low-carbon industry groups as "small, poorly organized, and politically weak," writing that "powerful coalitions of low-carbon industries are a topic for the future, not today."[18]

That future is arriving fast. In 2020, dozens of companies, including vehicle manufacturers, electric utilities, Uber, and charging providers, teamed up to form the Zero Emission Transportation Association, which advocates for 100 percent of vehicle sales to be electric by 2030. In 2021, trade groups for solar, storage, wind, and transmission companies allied with utilities and corporate purchasers of clean energy to launch the American Clean Power Association, which aims to "unit[e] the renewable power sector to speak with a unified voice." These groups are "a classic example of what policy folks would call policy feedback," Mildenberger told *The Atlantic*. "You create a new industry that creates a new set of interests that creates new bedfellows, and as that industry becomes more successful, it feeds back and increases the strength of the policy." The clout of low-carbon industries could continue to grow alongside their revenues as they seek to lock in or expand any advantages that policies may bestow upon them.[19]

Given all these changes in the political landscape, there is reason to hope fossil fuel corporations and their trade organizations will no longer obstruct climate legislation and clean energy rivals can more effectively advocate for progress. Still, like Frankenstein, the monster of climate denialism the fossil fuel industry created, through deceptions chronicled by Naomi Oreskes and Erik Conway in *Merchants of Doubt*, has taken on a life of its own. Reactionary politicians continue to oppose climate action, even as corporate interest groups have swung around to support it. The right-wing echo chamber of Fox News and Newsmax, Sinclair-owned local television stations, and conservative talk radio continues to distort climate science. Especially strident vitriol has been directed at women such as Rep. Alexandria Ocasio-Cortez (D-New York) and Greta Thunberg who champion climate action. When Rep. Matt Gaetz (R-Florida) issued a mild "Green Real Deal" as a conservative alternative to the Green New Deal, right-wing groups pounced, denouncing him as a "weak-minded congressional Republican" who had "fall[en] for the climate-baiting guilt-trip." One group sent a 16-page pamphlet to every Republican in Congress titled *The Plan is . . . No Plan! Why the GOP shouldn't do anything on climate*. "The Competitive Enterprise Institute and Heartland beat up on Gaetz like he's Al Gore, to make an example out of him," Taylor told me. "They essentially act as prison guards in the denialist penitentiary. Their job isn't to change anyone's mind, but to instill fear that if you change your mind, you'll be treated harshly." That's a powerful strategy in an era of gerrymandering, when Republican legislators fear primary challengers from the right more than Democratic opponents in the general election. Still, as public support for climate action grows and corporations withdraw funding from denialist groups, the effectiveness of this strategy could dwindle.[20]

Policy Pathways

With so many promising trends in diplomacy, technology, public opinion, and interest groups clearing away obstacles from policy pathways, which pathways should be pursued? No single pathway will suffice. Forging ahead toward a clean energy economy will require many actors pursuing various paths at once. Where advances are made, policies can mutually reinforce diplomacy and technology, making diplomacy more credible and technologies more affordable. Policy successes can also galvanize support from voters and interest groups, reshaping public opinion and fortifying ongoing pursuits in what political scientists call "policy feedback." As Jacob Hacker and Paul Pierson have written, "major public policies . . . can have substantial political impacts, engendering support that helps those policies to endure." Successes of initial policies strengthen the standing of clean energy industries and associated groups and build support to lock in and extend initial gains.[21]

Given the context of a Democratic administration and a closely divided Congress, I'll focus first on executive branch options before turning to legislation and then actions beyond the Beltway.

Regulations

Economists may dismiss regulations as less flexible and efficient than market-based approaches. But regulations get the job done. "Regulatory programs have generally delivered more change than markets, even though in theory markets could be more cost-efficient," Cullenward told me.[22]

The greatest opportunities to regulate greenhouse gases under existing law come from the Clean Air Act. "Without legislation, you're going to have to go through the keyhole of the Clean Air Act," said Joseph Goffman, who helped develop the Clean Power Plan before leading Harvard Law School's Environmental and Energy Law Program and then returning to EPA in the Biden administration. The simplest way to squeeze through that keyhole is to strengthen existing rules for emissions rates from vehicles, oil and gas operations, power plants, and other industries. Vehicle emission standards can be strengthened more readily now that many manufacturers are committing to produce electric cars. For industrial emitters, EPA can toughen its permitting and compliance efforts and file enforcement actions when violations are found. More vigorous enforcement or tightening of ambient air quality standards would inevitably lead to cuts in greenhouse gases that are co-emitted with other air pollutants. That would force states that have violated ozone smog or particulate matter standards with little consequence for decades to regulate polluting industries more stringently.[23]

For power plants, new source performance standards could be updated to mandate performance in line with the latest technologies. For example, the Allam cycle with carbon capture pioneered by NET Power can reset the benchmark for emissions from natural gas–fired electricity. Requiring old coal plants to install sulfur scrubbers, as has been required at all new plants since the early 1980s, would prompt many of them to close to avoid costly investments. Either the Obama-era Clean Power Plan or Trump-era Affordable Clean Energy rule could be reissued with much tougher emissions caps, made achievable by technological advances.

Section 115 of the Clean Air Act offers an untested path to regulating greenhouse gases. That section empowers EPA to regulate pollutants that endanger health or welfare in other countries, as greenhouse gases clearly do. For Section 115 to be applicable, the EPA administrator must determine that other countries are acting with reciprocity in controlling their own emissions. Paris Agreement commitments could satisfy that test. EPA could then order states to issue plans for curtailing their emissions, and it could issue a federal plan for states that do not comply. "This unutilized but potent source of federal authority provides an avenue to achieve significant greenhouse gas emissions reductions," one study of Section 115 concluded.[24]

However, this or other untested regulatory approaches risk being overturned by skeptical judges. As of 2021, Trump appointees occupy nearly one-third of appellate judgeships and three seats on the Supreme Court. Only one justice from the 5–4 majority in *Massachusetts v. EPA* and three of the dissenters, including Chief Justice John Roberts, remain on the Supreme Court. Justice Amy Coney Barrett clerked for Justice Antonin Scalia, who authored the scathing dissent in *Massachusetts v. EPA*. Even without an outright repeal by the courts, the years of delay for judicial review can give a new administration time to roll back a regulation, as the Trump administration did with the Clean Power Plan and other Obama-era rules. Thus, novelty is not a virtue for seeking urgently needed reductions in emissions.

Emissions Pricing

No climate policy has attracted more attention, enthusiasm, and derision than putting a price on emissions. The Waxman-Markey bill would have established such a price through a cap-and-trade system, with markets for emissions allowances. More recent bills have focused instead on taxing emissions. At least seven bills for carbon taxes were proposed in the 2019–2020 session of Congress, sponsored solely by Democrats and a lone retiring Republican. Each would have set a steadily rising, economy-wide tax on emissions, though the proposed rates varied widely. Most would have set a border adjustment tax on the carbon content of imports. Some would have suspended EPA regulations for stationary emission sources that are covered by the tax, to protect the competitiveness of American manufacturers against foreign rivals. Some of the bills would have returned most of the revenue to Americans as dividends or tax reductions, whereas others would have invested some of it in infrastructure, RD&D, or adaptation to climate change.[25]

The economic logic of emissions taxes is compelling. In fact, more than three thousand U.S. economists, including twenty-eight Nobel laureates, have endorsed the Climate Leadership Council's carbon tax and dividend plan. A National Academies panel on accelerating decarbonization recommended a similar tax rate, gradually rising from $40 per ton, but with revenue targeted to fund clean energy and address equity concerns rather than rebated entirely as a flat dividend. However the revenue is spent, taxing emissions addresses what economists call "externalities"—the health, climate, and other impacts borne by society rather than the consumers and producers of fossil fuels. Taxes shift some of that burden onto consumers or producers, motivating them to reduce emissions to avoid those taxes. Taxes on fossil fuels provide a market demand "pull" for everything from efficiency to clean energy to negative emissions technologies that avert taxation. "You'll see more technological progress than you expect," predicts Noah Kaufman, a carbon pricing expert serving on President Biden's Council of Economic Advisers.[26]

Designers of carbon taxes must decide what gets taxed and at what rate. Many economists prefer a uniform economy-wide tax, reflecting the uniform harm of carbon dioxide per ton. That would provide a broad and level playing field for anything that cuts emissions. Narrower policies may be criticized for favoring some options over others.

However, a tax rate that could drive meaningful change in one sector may do little in others. In the electric sector, for example, the $40 per ton starting point of the Climate Leadership Council plan would suffice to wipe out coal power plants, whose operating costs would more than double, and help renewables outcompete gas. Electricity consumers would see little change in their rates, now that renewables and their complements can affordably displace fossil electricity. For transportation, on the other hand, that same $40 per ton would add just 9 cents per liter (35 cents per gallon) to the cost of gasoline. That would barely influence decisions regarding which car to buy or how far to drive it. It could, however, trigger the visceral opposition that typically accompanies hikes in fuel prices, which are so prominently displayed at every fueling station. Opposition from drivers could sink the viability of an economy-wide carbon tax, negating the benefits it could achieve in other sectors.[27]

A uniform carbon tax would also fail to target the local harms that accompany emissions. Air pollution, water pollution, and drilling and mining wastes from fossil fuels all impose burdens that are not proportional to carbon dioxide and fall most heavily on vulnerable communities. A uniform tax also gives equal value to reducing emissions via efficiency, which averts all the harms of fossil fuels, and carbon capture, which addresses only the carbon dioxide.[28]

Beyond the scope of a tax comes the question of what to do with the revenue. Returning carbon tax revenue as dividends has been advocated mainly on the premise that it would win political support, both for the initial passage of a bill and for keeping it durable. "Once that system gets in place and people start getting checks, it will become a third rail of American politics to stop it," Leiserowitz told me, comparing carbon dividends to Social Security and the checks that Alaskans receive from oil and gas tax revenue. Dividends are seen as crucial by pro-climate Republicans such as Bob Inglis, the former congressman and executive director of RepublicEn, and Secretary Baker. "You're not ever going to get anything done if you don't bring Republicans to the table, and you do that by saying it's not going to grow government," Baker told me. Other dividend advocates point out the simplicity of avoiding battles among competing interests. "Once you start using revenue for other purposes, you open a Pandora's box," said Bertelsen of the Climate Leadership Council. Dividends also make a carbon tax progressive, with uniform dividends per person but more taxes paid by wealthy Americans with large homes and frequent air travel. Carbon taxes can be made even more progressive by means-testing dividends or using revenue to reduce payroll taxes, as some bills proposed to do.[29]

However, it is not clear that a dividend is the most effective or most popular use of carbon tax revenue. Rebating carbon tax revenue renders it unavailable to invest in RD&D or build out the transmission lines, vehicle chargers, green hydrogen facilities, carbon dioxide pipelines, and other infrastructure needed for decarbonization. It also fails to remedy environmental injustices in communities burdened by the legacy of fossil fuels. Surveys yield mixed results about voter preferences for carbon tax revenue, with most showing a preference for spending at least some revenue on clean energy RD&D, not just dividends. In Washington state, a revenue-neutral carbon tax and dividend ballot initiative was rejected by voters in 2016 by an even larger margin than a 2018 initiative that would have used the money to tackle climate change.[30]

All of these considerations about how to design a carbon tax are pointless if one cannot be passed. "Given the massive headwinds facing U.S. carbon taxes, debating their design kind of feels like debating what to have for lunch after we colonize Saturn," journalist Ben Geman has written. For all the talk from retired and retiring Republicans about dividends as the key to bipartisan support, few Republicans remaining in Congress have shown any willingness to accept carbon taxes, perhaps fearing attacks like those leveled at Representative Gaetz. A bipartisan group of centrist House lawmakers in 2021 floated the idea of hiking gas taxes rather than corporate taxes to fund infrastructure, but they faced opposition from colleagues in both parties. Tax-and-dividend has received only a mixed reception among congressional Democrats, appearing in some climate bills but not others.[31]

The United States is not alone in its failure to pass a carbon tax. But even in countries that have enacted carbon taxes, most are too low or too limited in scope to cut emissions deeply. That raises serious doubt whether a carbon tax can serve as the centerpiece of a push toward net zero.[32]

Social Cost of Greenhouse Gases

If robust emissions taxes can't be passed, the externalities of emissions can be considered administratively by setting social costs for carbon dioxide and other greenhouse gases. During the Obama era, an interagency working group recommended a valuation of roughly $50 per ton for carbon dioxide and much higher values for more potent greenhouse gases. The Trump administration disbanded that group, and it reset the value to just $1 to $6 for carbon dioxide by sharply discounting future effects and ignoring effects beyond U.S. borders. Just hours after taking office, President Biden issued an executive order that reestablished the working group. He ordered it not only to reassess social costs but also to recommend how these costs should be considered in decision-making, budgeting, and procurement across the federal government.[33]

Taken seriously, Biden's order could have profound implications. A social cost near $100 per ton, as some experts consider appropriate, would radically shift government purchasing decisions for vehicles, equipment, construction, and materials. This would create a powerful demand pull for decarbonization, especially if some state and local governments and corporations follow suit. The resulting learning by doing and technology diffusion curves could drive down costs for everyone else.

Beyond procurements, considering social costs in government decision-making could make it more difficult to justify approving mines, drilling sites, export terminals, and oil and gas pipelines that expand the use of fossil fuels. Projects like wind and solar farms, transmission lines, and hydrogen and carbon dioxide pipelines that avert emissions would be favored. Any state or local governments that follow suit would have their decision-making reshaped as well. Many corporations already consider an internal carbon price in their planning and investment decisions. Activist investors and interest groups can push for more to do so. Actions taken voluntarily beyond the federal government could stay durable as administrations come and go, all without a carbon tax to rally opposition.[34]

Green New Deal

Although a carbon tax has won the minds of economists, retired Republicans, and corporate coalitions, the Green New Deal has won the hearts of liberal activists. The idea of linking environmental protection with New Deal–scale job creation dates back at least to 2009, when Van Jones called for creating "green collar" jobs amid the Great Recession. The idea then leapt to prominence after the 2018 midterm elections, with the activism of groups like the Sunrise Movement and a resolution introduced by Representative Alexandria Ocasio-Cortez and Senator Edward Markey.[35]

As a mere 14-page resolution rather than detailed legislation, the Green New Deal would have enacted no binding policies, but its ambition was unfathomably broad. The bill set an overall aim of net-zero emissions along with specific calls for 100 percent clean electricity, efficiency upgrades for all buildings, "overhauling transportation systems," and collaborations with farmers and ranchers. Then, in a sweeping final clause, the resolution moved beyond climate policy to call for "providing all people of the United States with high-quality health care; affordable, safe, and adequate housing; economic security; and clean water, clean air, healthy and affordable food, and access to nature."[36]

With ambitions so broad and specifics so short, observers inevitably filled in the blanks to suit their agendas. The Sunrise Movement praised the Green New Deal as "the only plan put forward to address the interwoven crises of climate catastrophe, economic inequality, and racism at the scale that science and justice demand." President Trump warned that Democrats would "completely take over American energy and completely destroy America's economy through their new $100 trillion Green New Deal." Trump appeared to be rounding up from an unpublished estimate of a 10-year, $93 trillion cost, the vast majority of which would go to universal health care and guaranteed jobs, not climate. Seeking to score political points, Majority Leader Mitch McConnell (R-Kentucky) brought the Green New Deal up for a vote in the Senate, where it was rejected 57 to 0, with most Democrats abstaining. "The danger is that the Democratic party becomes so militant and uncompromising on the matter that Republicans decide they don't need to compromise, and they can just run against it," Taylor told me. Indeed, Republicans railed against the Green New Deal in the 2020 elections, and Joe Biden distanced himself from it while touting his own climate plan.[37]

It's easy to dismiss a resolution that failed so soundly and included no binding rules or funding. Yet the political insight behind the Green New Deal, and the reason it continues to inspire so many on the left and outpoll carbon taxes even among Republicans, is that it leads with the ends rather than the means. We all want clean air, clean water, a stable climate, and green jobs, even if we are not sure about the tax policy to achieve those goals. After decades of environmental wrongs, it is time to frame the climate battle in the context of environmental justice. Leading with what we want, rather than our means of getting it, inspires support beyond the realm of economists and policy wonks. "The key path forward to decarbonization is to move beyond discussing carbon and focus on the reasons to do it—health, resilience, diversification, security, jobs, competitiveness, and trade," environmental policy scholar Benjamin Sovacool told me.[38]

Infrastructure Investments

To transition from fossil fuels to clean energy and carbon sinks, we will need to build a lot of stuff. After years of partisan gridlock, the United States must regain its ability to build infrastructure. The American Society of Civil Engineers' 2021 Infrastructure Report Card rated the nation's energy systems a C– and noted a need for more transmission capacity to integrate renewable electricity. A National Academies panel on energy decarbonization in 2021 estimated that we will need $2 trillion of incremental capital investments in the next decade alone to get on track for net zero by 2050. Those are investments, not net costs, as most will pay for themselves with savings over time. Still, it's a lot of capital to mobilize very quickly. Most of the investments will come from the private sector to build wind and solar farms, buy zero-emissions vehicles and heat pumps, and improve the efficiency of buildings and industry. However, we will need big investments in shared infrastructure too—roughly $170 billion for electricity transmission and distribution, $80 billion for carbon dioxide pipelines and storage, and $6 billion for electric car chargers, above and beyond what would be spent in the status quo. The sooner those investments are made, the lower overall costs of decarbonization can be, since they become the framework into which other investments connect and avert expenses for fossil assets that are incompatible with a net-zero future.[39]

The challenges of building adequate infrastructure extend beyond cost. Transmission lines and carbon dioxide and hydrogen pipelines are all likely to face not-in-my-backyard opposition, mitigated somewhat by building them underground or along existing transportation and utility corridors to the extent possible. Government must streamline approval processes and ensure that infrastructure is built in a cost-effective, just, and timely manner. Incentives, low-interest loans, and other financing mechanisms can reduce the costs of construction.

Research, Development, and Deployment

Investments in research and development provide the technology "push" to make decarbonization affordable in the United States and around the world. "Every mandate, every policy should be tested against how is it driving technology, not just how is it moving emissions," said Severin Borenstein, an economist at the University of California, Berkeley. With the U.S. share of global emissions now at 13 percent and falling, making decarbonization more affordable and appealing beyond our borders is crucial. Unlike a border adjustment tax that burdens poor countries, technologies pushed by American RD&D and shared worldwide provide benefits for all.[40]

The United States is home to the most dynamic ecosystem for energy research through its national laboratories, universities, and other institutions and the talent they attract. The country also hosts the broadest array of public and private sources for funding the development and deployment of innovations that emerge. "Everything suggests that the U.S. economy is the crucible in which technologies are going to get forged and driven down in costs such that they're accessible to poor countries," said Adele Morris, an economist who studies climate policy at the Brookings Institution. "And that's how you solve climate change—you have to have cheap, low-emissions technologies so that the Chinas and Indias of the world can adopt them."[41]

Clean energy RD&D has for too long been underfunded, receiving just a tiny fraction of 1 percent of the federal budget. The National Academies panel on accelerating decarbonization has recommended tripling that funding, as have other experts. President Biden's climate plan calls for a $400-billion investment in clean energy research and innovation over a ten-year span.[42]

A balanced research portfolio should contain a mix of high-risk, high-reward pursuits in the style of ARPA-E programs, as well as cleantech incubator labs and funding to help developers bring their innovations to market. Priority should go to pivotal technologies where the United States is positioned to lead. Those include next-generation solar technologies like perovskites and solar organics; advanced biofuels; storage technologies; and methods to enhance the reliability of electricity supply and the flexibility of demand. The United States is especially well positioned to lead in advanced geothermal technologies, given its leadership in horizontal drilling technologies and pilot projects underway. Idled equipment and expertise from the oil and gas industries can be redirected toward geothermal and carbon capture. The United States has lagged Europe in pursuing clean hydrogen, but could boost its efforts if DOE's Hydrogen Shot initiative is adequately funded.[43]

The United States should also take the lead in RD&D for certain carbon sinks. Abundant agricultural land and expertise from land-grant colleges position the United States to pioneer better methods for sequestering carbon in soils and reducing nitrogen emissions from agriculture. Our extensive pipeline networks and depleted oil and gas reservoirs can be repurposed for carbon capture and storage. Geological conditions in portions of the western United States allow for testing of accelerated weathering and alkalinity-based approaches to removing carbon from the air.

Supply-Side Policies

Restrictions on fossil fuel supply do not attract nearly as much scholarly attention as carbon pricing, but they excite more passion from activists. Keep-it-in-the-ground protests against coal mines, oil fields, and pipelines shift the window of policy possibilities. "The anti–fossil fuel movement has shifted the politics and discourse leftward, which increases the salience of carbon tax discussions and makes a tax look more reasonable," said Fergus Green, a political scientist at Utrecht University.[44]

President Obama applied a keep-it-in-the-ground rationale to block the Keystone XL pipeline and set a moratorium for new coal mining leases on federal land. Both policies were reversed by President Trump and then reinstated by President Biden. Accounting for the social costs of greenhouse gases could make it easier to block leases for fossil fuel extraction on federal land and permits for oil and gas pipelines and coal export terminals.[45]

Whether to ban hydraulic fracturing (fracking) for oil and gas was a hot button topic in the 2020 presidential debates. However, most of the damage related to oil and gas comes from other steps in the extraction process, along with transporting, refining, and burning the fuels, not from fracking itself. Thus, while it is important to reduce supply and demand of all fossil fuels, singling out one particular extraction method is not the most prudent approach.

One danger of supply-side activism is that it can overreach into NIMBY (not in my backyard) or even BANANA (build absolutely nothing anywhere near anyone) opposition to all energy infrastructure. Decarbonizing energy will require building many things—wind and solar farms, geothermal wells, transmission lines, electric vehicle charging stations, hydrogen and carbon dioxide pipelines, carbon dioxide removal sites, and so on. Wind farms and transmission lines have too often been stalled or blocked by NIMBY or BANANA objections. Accounting for social costs of greenhouse gases can help distinguish what projects should be built.

Litigation

While supply-side policies draw skepticism from some climate policy scholars and economists, litigation can draw outright scorn. However, it was lawsuits that held the tobacco industry accountable for the hazards of smoking, and pharmaceutical companies accountable for the opioid epidemic. Books such as *Merchants of Doubt* have described the parallel historical deceptions of the fossil fuel and tobacco industries.

At the time of this writing, no lawsuit has won a major payout from an energy company for climate damage, although a Dutch court in 2021 ordered Shell to cut its emissions. That record makes it easy to dismiss the importance of the hundreds of cases that have been filed on behalf of youths, local governments, and others. Still, fear of litigation may be motivating corporate support for climate legislation, with hopes that it would be paired with legal indemnification. In fact, the industry-backed Climate Leadership Council's carbon dividends plan argues that "robust carbon taxes would also make possible an end to federal and state tort liability for emitters." Although some environmentalists oppose a liability shield, if it takes a shield to smooth the path toward robust legislation, preferably stronger than the Council's plan, that's a deal worth taking.[46]

State and Local Policies

Federal progress on climate policies ground to a halt, or in some cases fell backward, during the Trump era. Yet those four years saw an unprecedented flowering of activity by states and cities. Before President Trump took office, only Hawaii had committed to 100 percent clean electricity. By late 2019, ten other states plus the District of Columbia and Puerto Rico had done so too. The Sierra Club tracks more than 160 cities that have committed to 100 percent clean energy. The We Are Still In coalition tallies 10 states and 190 cities and counties that signed a declaration to uphold the aims of the Paris Agreement after President Trump announced plans to withdraw.[47]

State-level ambition has tended to swing as a counterweight to federal efforts. Republicans gained hundreds of seats in state legislatures amid a Tea Party backlash to President Obama. State-level progress slowed, and state attorneys general sued to block Obama administration regulations such as the Clean Power Plan. Democrats won back hundreds of seats in the 2018 midterm elections amid a backlash to President Trump. The number of states with Democratic "trifectas"—Democrats controlling the governor's office and both houses of the legislature—jumped from nine to fifteen, while Republicans lost four of their trifectas that year.[48]

Democratic trifecta states have pioneered the most ambitious state-level policies. For example, New York has enacted ambitious policies to improve building efficiency, electrify heating, and reduce vehicle emissions. Washington has created its own Clean Energy Fund to sponsor clean technology R&D. California has implemented an astounding array of policies including emissions trading markets, net-zero electricity building codes, and low-carbon fuels standards, and it will require all new heavy-duty trucks to be zero-emissions by 2045. These states and others are serving as laboratories of innovation, experimenting with new approaches in what some scholars describe as "experimentalist governance." It will be important to objectively evaluate the outcomes of those experiments so that successes can become the basis for adoption elsewhere. To foster this process, the National Academies panel recommended funding for policy evaluation studies and regional innovation hubs.[49]

Still, state and local actions are no substitute for a coordinated federal response. For diplomacy, only the federal government can issue a national commitment under the Paris Agreement or drive the creation of climate clubs. For technology, research motivated by state-level electricity and vehicle mandates has nowhere near the firepower that would come with federal mandates and investments. State-level zero-emissions vehicle requirements may not lead to the build-out of electric charging or hydrogen refueling infrastructure needed for interstate driving, especially if truckers buy their trucks in states with laxer rules. State policies are also vulnerable to "leakage," if tough regulations or carbon prices in one state lead industries to move their manufacturing to others.

Putting It All Together

With a more promising political landscape than ever before and so many policy pathways branching ahead, which ones should be pursued?

Net zero by 2050 sets a framing for other pursuits. That target should be enshrined in our national commitment under the Paris Agreement, replacing our initial aim for an 80 percent reduction in greenhouse gases. The European Union and a growing list of countries have already set such targets, and an American commitment can help encourage others to follow suit.

More specific commitments will be needed for key sectors, including targets for clean electricity, zero-emissions vehicles, net-zero emissions new buildings, and carbon dioxide removal. Clean electricity should be pursued most rapidly, since it is pivotal to enhancing efficiency, making electrification worthwhile, and powering negative emissions technologies. Setting a standard along the lines of 90 percent clean electricity by 2035 would ensure a rapid transformation while keeping electricity affordable and reliable as other sectors continue to electrify. Regulators should streamline approvals of the transmission lines, onshore and offshore wind and solar sites, geothermal drilling, and nuclear permit extensions needed to achieve that electricity standard. Congress and states should issue a mix of regulations and incentives to accelerate the electrification of vehicles, heating, and industries to use that clean electricity.

All these policies and associated investments to decarbonize energy should be pursued as quickly as possible. By contrast, carbon dioxide removal from the atmosphere, which depends on clean electricity and still unproven techniques, merits more research than deployment for now.

Ideally, a steadily rising emissions tax would provide a market signal to stimulate the pursuit of net zero. It would incentivize both emissions abatement and removal, and it would help companies plan prudently for investments and retirements of assets. Such a tax should extend beyond carbon dioxide to cover other greenhouse gases and air pollutants, enhancing air quality alongside climate. If it were up to me, some of the revenue would be devoted to means-tested dividends for progressivity, some to communities affected by climate change and the energy transition, and some to clean energy RD&D and infrastructure.

But economists, policy wonks, retired Republicans, and I do not make policy. At the time of this writing, even with a Democratic administration, it is hard to imagine sixty senators supporting a carbon tax at a level sufficient to drive decarbonization. A tax is so hard to pass because it sets controversial means, rather than widely sought ends, at the heart of the policy. Targets for clean electricity, mandates for efficiency, and investments in RD&D and infrastructure all set the ends ahead of the means, even if they are not as theoretically optimal. We don't have time to wait for perfect policies.

As the old adage from Otto von Bismarck puts it, "Politics is the art of the possible, the attainable—the art of the next best." If a sufficiently high carbon price cannot pass, a strategy must be patched together from what can.

Mandates will play a role. When we have truly wanted to get rid of things—ozone-depleting CFCs, leaded gasoline, uncontrolled mercury emissions—we have not done it by taxing them; we have banned them. Coal-fired electricity, petroleum-fueled vehicles, and HFCs all have substitutes that can allow for similar phase-outs. A growing number of cities, states, and countries are already mandating phase-outs for each of those items. Major utilities and automobile manufacturers are setting their own timetables for phase-outs of fossil fueled electricity and vehicles.

Policies for reducing fossil fuel demand and emissions should be accompanied by constraints on supply too. That includes banning new coal mines and the export of coal, restricting oil and gas drilling near populated areas and sensitive ecosystems, and rejecting new oil and gas pipelines except where needed to avert methane leaks and flaring. Vigorous provisions should be enacted to avoid methane leaks, oil spills, and environmental damage from the oil and gas extraction and transport that remain. Litigation should provide a backstop to ensure adherence to these and other regulations.

In the international arena, the United States should incorporate its policies into its Paris commitment and push hard diplomatically to persuade other countries to ratchet up their own commitments. That push should be bolstered by climate clubs, offering membership and trading privileges only to countries that issue and remain on track for ambitious commitments. International cooperation should facilitate RD&D of clean energy technologies and ensure that they are readily shared to drive down costs around the world.

Of course, not all of this will be achieved. Duplication and redundancy will be a feature, not a bug, of climate policies, so that the push toward decarbonization can power forward even if a few policies are reversed. Actions by states, cities, and companies provide crucial backstops if federal action lags, and multiple arenas of action offer the experimentation needed for best practices to emerge. Success will come not from a single piece of legislation but from a continual ratcheting up of efforts to stimulate new technologies, invest in infrastructure, and leverage U.S. actions via diplomacy.

Time is of the essence. The longer it takes the world to reach net zero, the hotter our climate will become. The United States has a pivotal role to play in that pursuit. We are already too late to avert substantial warming, and perhaps even to hold warming below 1.5°C. But with concerted and persistent efforts, we can indeed confront climate gridlock, unlock a clean energy future, and help our children and grandchildren inherit a more hospitable Earth.

NOTES

Chapter 1. Why Climate Gridlock?

1. The phrase "global warming gridlock" is from David Victor, *Global Warming Gridlock* (Cambridge: Cambridge University Press, 2011).

2. The total number of ratifications was 189 as of July 9, 2020, according to the United Nations, https://cop23.unfccc.int/process/the-paris-agreement/status-ofss-ratification. Greenhouse gas emissions percentages are for 2018 from Table B.1 of J.G.J. Olivier and J.A.H.W. Peters, "Trends in Global CO2 and Total Greenhouse Gas Emissions: 2019 Report," May 2020, 70.

3. Olivier and Peters, "Trends in Global CO2 and Total Greenhouse Gas Emissions."

4. Varun Sivaram, Zoom interview, April 6, 2020.

5. D. Gilfillan et al., "Global, Regional, and National Fossil-Fuel CO2 Emissions: 1751–2016" (Boone, N.C.: Carbon Dioxide Information Analysis Center at Appalachian State University, 2019), https://energy.appstate.edu/CDIAC. Pierre Friedlingstein et al., "Global Carbon Budget 2019," *Earth System Science Data* 11, no. 4 (December 4, 2019): 1783–1838, https://doi.org/10.5194/essd-11-1783-2019. Andrew P. Schurer et al., "Importance of the Pre-Industrial Baseline for Likelihood of Exceeding Paris Goals," *Nature Climate Change* 7, no. 8 (August 2017): 563–67, https://doi.org/10.1038/nclimate3345. Ed Hawkins et al., "Estimating Changes in Global Temperature Since the Preindustrial Period," *Bulletin of the American Meteorological Society* 98, no. 9 (September 2017): 1841–56, https://doi.org/10.1175/BAMS-D-16-0007.1.

6. NOAA National Centers for Environmental Information, Climate at a Glance: Global Time Series, published July 2020, retrieved on July 17, 2020, from www.ncdc.noaa.gov/cag. Chris Mooney, Andrew Freedman, and John Muyskens, "2020 Rivals Hottest Year on Record, Pushing Earth Closer to a Critical Climate Threshold," *Washington Post*, January 14, 2021, www.washingtonpost.com/climate-environment/interactive/2021/2020-tied-for-hottest-year-on-record. Robert Rohde, "Global Temperature Report for 2020" (Berkeley Earth, January 14, 2021), http://berkeleyearth.org/global-temperature-report-for-2020.

7. IPCC, "Global Warming of 1.5°C. An IPCC Special Report on the Impacts of Global Warming of 1.5°C above Pre-Industrial Levels and

Related Global Greenhouse Gas Emission Pathways," 2018, www.ipcc.ch
/sr15.

8. Richard J. Millar and Pierre Friedlingstein, "The Utility of the Historical
Record for Assessing the Transient Climate Response to Cumulative
Emissions," *Philosophical Transactions of the Royal Society A: Mathematical,
Physical, and Engineering Sciences* 376, no. 2119 (May 13, 2018): 20160449,
https://doi.org/10.1098/rsta.2016.0449. Richard J. Millar et al., "Emission Budgets and Pathways Consistent with Limiting Warming to 1.5 °C,"
Nature Geoscience 10, no. 10 (October 2017): 741–47, https://doi.
org/10.1038/ngeo3031. H. Damon Matthews et al., "Estimating Carbon
Budgets for Ambitious Climate Targets," *Current Climate Change Reports*
3, no. 1 (March 1, 2017): 69–77, https://doi.org/10.1007/s40641-017-0055-
0. "Analysis: How Much 'Carbon Budget' Is Left to Limit Global
Warming to 1.5C?," *Carbon Brief*, April 9, 2018, www.carbonbrief.org/
analysis-how-much-carbon-budget-is-left-to-limit-global-warming-
to-1-5c.

A barrel of oil is 0.43 tons CO_2, according to the U.S. Environmental
Protection Agency; EPA, "Greenhouse Gases Equivalencies Calculator—
Calculations and References | Energy and the Environment," accessed
July 18, 2020, www.epa.gov/energy/greenhouse-gases-equivalencies-
calculator-calculations-and-references.

9. The Hurricane Harvey likelihood is from Kerry Emanuel, "Assessing the
Present and Future Probability of Hurricane Harvey's Rainfall," *Proceedings of the National Academy of Sciences* 114, no. 48 (November 28, 2017):
12681, https://doi.org/10.1073/pnas.1716222114, and Geert Jan van Olden-
borgh et al., "Attribution of Extreme Rainfall from Hurricane Harvey,
August 2017," *Environmental Research Letters* 12, no. 12 (December 2017):
124009, https://doi.org/10.1088/1748-9326/aa9ef2.

David J. Frame et al., "The Economic Costs of Hurricane Harvey Attributable to Climate Change," *Climatic Change* 160, no. 2 (May 1, 2020):
271–81, https://doi.org/10.1007/s10584-020-02692-8.

Rain rates are from Figure 2.6 of USGCRP, "Impacts, Risks, and Adaptation in the United States: Fourth National Climate Assessment," 2018,
https://nca2018.globalchange.gov. The change in Arctic sea ice is taken
from Figure 2.7 of USGCRP. Greenland ice melt is drawn from Chapter
2 of USGCRP and converted to centimeters based on 1 km³ per gigaton of
water and 2.166 million km² surface area of Greenland. This and subsequent mentions of "tons" refer to metric tons.

Melting of glaciers is from IPCC, "Global Warming of 1.5°C. An IPCC
Special Report on the Impacts of Global Warming of 1.5°C Above

Pre-Industrial Levels and Related Global Greenhouse Gas Emission Pathways."

10. IPCC, "Global Warming of 1.5°C."

11. IPCC, "Global Warming of 1.5°C." C.-F. Schleussner et al., "Differential Climate Impacts for Policy-Relevant Limits to Global Warming: The Case of 1.5 °C and 2 °C," *Earth System Dynamics Discussions* 6, no. 2 (November 27, 2015): 2447–2505, https://doi.org/10.5194/esdd-6-2447-2015. Peter U. Clark et al., "Sea-Level Commitment as a Gauge for Climate Policy," *Nature Climate Change* 8, no. 8 (August 2018): 653–55, https://doi.org/10.1038/s41558-018-0226-6.

12. Solomon M. Hsiang, Marshall Burke, and Edward Miguel, "Quantifying the Influence of Climate on Human Conflict," *Science* 341, no. 6151 (September 13, 2013), https://doi.org/10.1126/science.1235367. Marshall Burke, Solomon M. Hsiang, and Edward Miguel, "Global Non-Linear Effect of Temperature on Economic Production," *Nature* 527, no. 7577 (November 2015): 235–39, https://doi.org/10.1038/nature15725. Solomon Hsiang et al., "Estimating Economic Damage from Climate Change in the United States," *Science* 356, no. 6345 (June 30, 2017): 1362–69, https://doi.org/10.1126/science.aal4369. USGCRP, "Impacts, Risks, and Adaptation."

13. T. M. Lenton et al., "Tipping Elements in the Earth's Climate System," *Proceedings of the National Academy of Sciences* 105, no. 6 (February 12, 2008): 1786–93, https://doi.org/10.1073/pnas.0705414105. Timothy M. Lenton et al., "Climate Tipping Points—Too Risky to Bet Against," *Nature* 575, no. 7784 (November 2019): 592–95, https://doi.org/10.1038/d41586-019-03595-0. IPCC, "Global Warming of 1.5°C." Timothy M. Lenton, Phone interview, September 13, 2019. Will Steffen et al., "Trajectories of the Earth System in the Anthropocene," *Proceedings of the National Academy of Sciences* 115, no. 33 (August 14, 2018): 8252–59, https://doi.org/10.1073/pnas.1810141115.

14. "Temperatures | Climate Action Tracker," accessed July 12, 2020, https://climateactiontracker.org/global/temperatures.

    J. A. Church, P. U. Clark, and A. Cazenave, "Sea Level Change," in *Climate Change 2013: The Physical Science Basis. Contribution of Working Group I to the Fifth Assessment Report of the Intergovernmental Panel on Climate Change*, ed. T. F. Stocker and D. Qin (Cambridge: Cambridge University Press, 2013), Chapter 13. IPCC, "Global Warming of 1.5°C." Katharine Hayhoe, Phone interview, June 11, 2018. G. Naumann et al., "Global Changes in Drought Conditions Under Different Levels of Warming," *Geophysical Research Letters* 45, no. 7 (2018): 3285–96, https://doi.org/10.1002/2017GL076521.

15. IPCC, "Global Warming of 1.5°C." Kelly Levin et al., "Designing and Communicating Net-Zero Targets" (World Resources Institute, July 2020), www.wri.org/design-net-zero. United Nations Environment Programme, "Emissions Gap Report 2020" (Nairobi, December 1, 2020), www.unep.org/emissions-gap-report-2020.

16. The Zero Carbon Consortium, "America's Zero Carbon Action Plan" (United Nations Sustainable Development Solutions Network, 2020), https://irp-cdn.multiscreensite.com/6f2c9f57/files/uploaded/zero-carbon-action-plan%20%281%29.pdf. E. Larson et al., "Net-Zero America: Potential Pathways, Infrastructure, and Impacts, Interim Report" (Princeton: Princeton University, December 15, 2020). National Academies of Sciences, Engineering, and Medicine, *Accelerating Decarbonization of the U.S. Energy System* (Washington, D.C.: National Academies Press, 2021), https://doi.org/10.17226/25932. "A Policy Pathway to Reach U.S. Net Zero Emissions by 2050" (Energy Innovation, November 15, 2019), https://energyinnovation.org/publication/a-policy-pathway-to-reach-u-s-net-zero-emissions-by-2050. "Getting to Zero: A U.S. Climate Agenda" (Center for Climate and Energy Solutions, November 8, 2019), www.c2es.org/getting-to-zero-a-u-s-climate-agenda-report/introduction.

17. Vaclav Smil, *Energy and Civilization: A History* (Cambridge: MIT Press, 2017). Matto Mildenberger, Phone interview, December 10, 2018. Rafe Pomerance, Phone interview, June 20, 2019.

Chapter 2. The Road to Paris

1. William Sweet, *Climate Diplomacy from Rio to Paris: The Effort to Contain Global Warming* (New Haven: Yale University Press, 2016), 177, 182. UN News Center, "UN Chief Hails New Climate Change Agreement as 'Monumental Triumph,' " *United Nations Sustainable Development* (blog), December 12, 2015, www.un.org/sustainabledevelopment/blog/2015/12/un-chief-hails-new-climate-change-agreement-as-monumental-triumph. "Obama: Climate Deal Is 'Best Chance We Have to Save the One Planet We've Got,' " *NBC News*, December 12, 2015, www.nbcnews.com/news/us-news/obama-climate-deal-best-chance-we-have-save-one-planet-n479026.

2. Donald Trump, "Statement by President Trump on the Paris Climate Accord," White House, June 1, 2017, www.whitehouse.gov/briefings-statements/statement-president-trump-paris-climate-accord.

3. Mario J. Molina and F. S. Rowland, "Stratospheric Sink for Chlorofluoromethanes: Chlorine Atom-Catalysed Destruction of Ozone," *Nature* 249, no. 5460 (June 1974): 810–12, https://doi.org/10.1038/249810a0.

"The Nobel Prize in Chemistry 1995," NobelPrize.org, accessed July 12, 2020, www.nobelprize.org/prizes/chemistry/1995/summary. The Nobel Prize was shared by Mario Molina, F. Sherwood Rowland, and Paul Crutzen for "their work in atmospheric chemistry, particularly concerning the formation and decomposition of ozone." National Research Council, *Halocarbons: Effects on Stratospheric Ozone*, 1976, https://doi.org/10.17226/19978. Toxic Substance Control Act, Pub. L. No. 94-469, 90 Stat. 2003 (1976). Peter M. Morrisette, "The Evolution of Policy Responses to Stratospheric Ozone Depletion," *Natural Resources Journal* 29 (1989): 793–820. Note: the EPA's ban on non-essential CFCs in aerosols was proposed in 1977 and took effect in 1978.

4. The British team's finding was announced at Ozone Commission meeting in Halkidiki, Greece in September 1984 and later published as J. C. Farman, B. G. Gardiner, and J. D. Shanklin, "Large Losses of Total Ozone in Antarctica Reveal Seasonal ClOx /NOx Interaction," *Nature* 315, no. 6016 (May 1985): 207–10, https://doi.org/10.1038/315207a0. These results were confirmed by NASA in R. S. Stolarski et al., "Nimbus 7 Satellite Measurements of the Springtime Antarctic Ozone Decrease," *Nature* 322, no. 6082 (August 1986): 808–11, https://doi.org/10.1038/322808a0.

5. Farman, Gardiner, and Shanklin, "Large Losses of Total Ozone in Antarctica Reveal Seasonal ClOx /NOx Interaction." Stolarski et al., "Nimbus 7 Satellite Measurements of the Springtime Antarctic Ozone Decrease."

Vienna Convention for the Protection of the Ozone Layer, 1985. The "Most successful treaty" quote is from Stephen Leahy, "Without the Ozone Treaty You'd Get Sunburned in 5 Minutes," *National Geographic*, September 25, 2017, www.nationalgeographic.com/news/2017/09/montreal-protocol-ozone-treaty-30-climate-change-hcfs-hfcs. EPA estimates of impacts are taken from Environmental Protection Agency, "Updating Ozone Calculations and Emissions Profiles for Use in the Atmospheric and Health Effects Framework Model" (2015), www.epa.gov/sites/production/files/2015-07/documents/updating_ozone_calculations_and_emissions_profiles_for_use_in_the_atmospheric_and_health_effects_framework_model.pdf. "Eventually" reflects the fact that this is over the first two centuries of the protocol.

6. Clean Air Act provisions can be found at www.epa.gov/ozone-layer-protection/ozone-protection-under-title-vi-clean-air-act. Chemical industry strategy is based on Brigitte Smith, "Ethics of Du Pont's CFC Strategy, 1975–1995," *Journal of Business Ethics*, 1998, 13.

7. Smith, "Ethics of Du Pont's CFC Strategy."

8. Pierre Friedlingstein et al., "Carbon Budget and Trends 2019" (Global Carbon Project, December 4, 2019), www.globalcarbonproject.org/car bonbudget. U.S. data are taken from EIA, "Fossil Fuels Still Dominate U.S. Energy Consumption Despite Recent Market Share Decline— Today in Energy—U.S. Energy Information Administration (EIA)," July 1, 2016, www.eia.gov/todayinenergy/detail.php?id=26912. Global data are taken from "Fossil Fuel Energy Consumption (% of Total) | Data," accessed July 13, 2020, https://data.worldbank.org/indicator/EG.USE. COMM.FO.ZS. Susan Biniaz, Phone interview, June 28, 2019.

9. National Research Council, *Carbon Dioxide and Climate: A Scientific Assessment* (Washington, D.C.: National Academies Press, 1979), https://doi. org/10.17226/12181. Ronald Reagan and Mikhail Gorbachev, "Joint Statement Following the Soviet-United States Summit Meeting in Moscow," June 1, 1988, www.reaganlibrary.gov/research/speeches/060188b. F. Kenneth Hare, "World Conference on the Changing Atmosphere: Implications for Security, Held at the Toronto Convention Centre, Toronto, Ontario, Canada, during 27–30 June 1988," *Environmental Conservation* 15, no. 3 (1988): 282–83, https://doi.org/10.1017/S0376892900029635. "Global Warming Prevention Act of 1988 (1988—H.R. 5460)," Pub. L. No. H.R. 5460 (1988), www.govtrack.us/congress/bills/100/hr5460. Timothy Wirth, "S.2667— 100th Congress (1987–1988): National Energy Policy Act of 1988" (1988), www.congress.gov/bill/100th-congress/senate-bill/2667. Though a record warm year globally at the time, 1988 no longer ranks in the top 25. In fact, there have been only 4 cooler years since, according to data from NOAA. The 43 days above 90 degrees in Washington in July–August 1988 was a record at the time, not surpassed until 2016. Jason Samenow, "D.C. Hits 90 Degrees for 50th Time This Year; Most 90s on Record in July and August," *Washington Post*, August 30, 2016, www.washingtonpost.com/news/capital-weather-gang/wp/2016/08/30/d-c-hits-90-degrees-for-50th-time-this-year-most-90s-on-record-in-july-and-august. President Bush quotes from *McNeil/Lehrer NewsHour*, September 1, 1988, accessed from Lexis Uni. and "Opinion | The White House and the Greenhouse," *New York Times*, May 9, 1989, sec. Opinion, www.nytimes.com/1989/05/09/opinion/the-white-house-and-the-greenhouse.html.

10. Nathaniel Rich, *Losing Earth: A Recent History* (New York: MCD/Farrar, Straus and Giroux, 2019), Chapters 18, 19, and 21. Fox Butterfield, "Sununu's Role in Debates on Environmental Issues Reflects Old Patterns," *New York Times*, May 14, 1990, sec. U.S., www.nytimes.com/1990/05/14/ us/sununu-s-role-in-debates-on-environmental-issues-reflects-old-pat terns.html.

11. James Morton Turner and Andrew C. Isenberg, *The Republican Reversal: Conservatives and the Environment from Nixon to Trump* (Cambridge: Harvard University Press, 2018), 156. Joshua Busby, "A Green Giant? Inconsistency and American Environmental Diplomacy," in *America, China, and the Struggle for World Order: Ideas, Traditions, Historical Legacies, and Global Visions*, ed. J. G. John Ikenberry, Zhu Feng, and Wang Jisi, Asia Today (Palgrave Macmillan US, 2015), https://doi.org/10.1057/9781137508317. Elizabeth DeSombre, "The United States and Global Environmental Politics: Domestic Sources of U.S. Unilateralism," in *The Global Environment: Institutions, Law, and Policy*, ed. Regina Axelrod and David Leonard Downie (Washington, D.C.: CQ Press, 2005), 192–212.

12. The developed nations' target is taken from David Wirth and Daniel Lashof, "Beyond Vienna and Montreal: A Global Framework Convention on Greenhouse Gases," *Transnational Law and Contemporary Problems* 2, no. 1 (1992): 90–91. Only Turkey joined the United States in resisting this target. IPCC, "Climate Change: The IPCC 1990 and 1992 Assessments," 1992, www.ipcc.ch/report/climate-change-the-ipcc-1990-and-1992-assessments.

13. The phrase "bet your economy decision" comes from Rich, *Losing Earth: A Recent History*, 178, clarified by email communication with Nathaniel Rich, May 21, 2019. President Bush's threat is in Geoffrey Palmer, "The Earth Summit: What Went Wrong at Rio?," *Washington University Law Quarterly* 70, no. 4 (1992): 1005–28. Discussion of the Rio treaty is based on Kyle Danish, "The International Regime," in *Global Climate Change and U.S. Law*, ed. Michael Gerrard (American Bar Association, 2007), 10–15.

14. The quotation "constitution for international action on climate change" is taken from Danish, "The International Regime."

15. For requirements for developing countries, see United Nations Framework Convention on Climate Change, 1992, Article 3. The role of President Bush is examined in Palmer, "The Earth Summit: What Went Wrong at Rio?"

16. The Rio treaty and Vienna Convention are compared in Sweet, *Climate Diplomacy from Rio to Paris*, 127–28. London Amendments to Montreal Protocol are discussed in Joel A. Mintz, "Progress Toward a Healthy Sky: An Assessment of the London Amendments to the Montreal Protocol on Substances That Deplete the Ozone Layer," *Yale Journal of International Law* 16 (1991): 571–82. Trends in CFCs and stratospheric ozone are assessed in WMO/UN Environment Programme, Scientific Assessment of Ozone Depletion: 2018.

The ozone hole in 2019 was the smallest since 1982, according to NASA: Sara Blumberg, "2019 Ozone Hole Is the Smallest on Record Since Its Discovery," NASA, October 21, 2019, www.nasa.gov/feature/goddard/2019/2019-ozone-hole-is-the-smallest-on-record-since-its-discovery.

17. The trend in emissions is based on Olivier and Peters, "Trends in Global CO2 and Total Greenhouse Gas Emissions: 2019 Report." The trend in temperatures comes from NOAA National Centers for Environmental Information, Climate at a Glance: Global Time Series, published June 2019, retrieved on June 18, 2019, from www.ncdc.noaa.gov/cag.

18. Thresholds for ratifying treaties are discussed by Andrew Moravcsik, "The Paradox of US Human Rights Policy," in *American Exceptionalism and Human Rights*, ed. Michael Ignatieff (Princeton: Princeton University Press, 2005), 147–97.

19. The Clinton and Bush positions are discussed by Shardul Agrawala and Steinar Andresen, "Indispensability and Indefensibility? The United States in the Climate Treaty Negotiations," *Global Governance* 5, no. 4 (1999): 457–82.

Developing country opposition is discussed by Daniel Bodansky, "The Copenhagen Climate Change Conference: A Postmortem," *American Journal of International Law* 104, no. 2 (2010): 230–40, https://doi.org/10.5305/amerjintelaw.104.2.0230.

20. Berlin talks are discussed by Sweet, *Climate Diplomacy from Rio to Paris*, 129, and by Agrawala and Andresen, "Indispensability and Indefensibility?" The Murkowski quote is from a press release, "U.S. Gets Hoodwinked at Berlin Climate Convention," issued by the Senate Committee on Energy and Natural Resources, April 7, 1995.

21. The U.S. position is described by AP, "U.S. Calls for Binding Action to Combat Climate Change," *The Columbian*, July 17, 1996.

The Murkowski quote is from David Fish, "U.N. Climate Change Negotiations in Geneva: Playing Poker with America's Economic Future, Says Murkowski," Senate Committee on Energy and Natural Resources, July 18, 1996.

22. For the Byrd-Hagel Senate resolution, see Robert C. Byrd, "S.Res.98—105th Congress (1997–1998): A Resolution Expressing the Sense of the Senate Regarding the Conditions for the United States Becoming a Signatory to Any International Agreement on Greenhouse Gas Emissions under the United Nations Framework Convention on Climate Change" (1997), www.congress.gov/bill/105th-congress/senate-resolution/98.

The quotes from senators are taken from "Senate Debate over Byrd-Hagel Resolution," *Congressional Record* 143, no. 107 (July 25, 1997).

23. The developing countries' position is discussed by Joshua Busby and Nigel Purvis, "Climate Leadership in Uncertain Times" (Atlantic Council Global Energy Center, September 11, 2018). Emissions per capita data for 1997 are from World Bank data. Melinda Kimble, Phone interview, June 7, 2019.

24. Sweet, *Climate Diplomacy from Rio to Paris*, 131.

25. Sweet, *Climate Diplomacy from Rio to Paris*, 132. Emma Paulsson, "A Review of the CDM Literature: From Fine-Tuning to Critical Scrutiny?," *International Environmental Agreements: Politics, Law & Economics* 9, no. 1 (March 2009): 63–80, https://doi.org/10.1007/s10784-008-9088-0.

26. Kimble, interview. Turner and Isenberg, *The Republican Reversal*, 167.

27. Nigel Purvis, Phone interview, May 22, 2019. Pomerance, interview.

28. The Bush campaign position is from Turner and Isenberg, *The Republican Reversal*, 168–69. The role of Powell is based on my interview with Nigel Purvis. O'Neill's role is based on Chuck Sudetic, "Bush's CO2 Flip-Flop: The Surprising Truth," *Rolling Stone*, May 10, 2001, www.rollingstone.com/politics/politics-news/bushs-co2-flip-flop-the-surprising-truth-55333. The Senate grand bargain is discussed by Michaël Aklin and Matto Mildenberger, "Prisoners of the Wrong Dilemma: Why Distributive Conflict, Not Collective Action, Characterizes the Politics of Climate Change," SSRN Scholarly Paper (Rochester, NY: Social Science Research Network, November 8, 2018), https://doi.org/10.2139/ssrn.3281045.

29. Greg Kahn, "The Fate of the Kyoto Protocol Under the Bush Administration," *Berkeley Journal of International Law* 21, no. 3 (2003): 25. For the alarm expressed by the fossil fuel industry, see Sudetic, "Bush's CO2 Flip-Flop."

   Lobbyist interactions with senators are discussed by Turner and Isenberg, *The Republican Reversal*, 170. George W. Bush, "Letter to Members of the Senate on the Kyoto Protocol on Climate Change," March 13, 2001. Lou Dubose and Jake Bernstein, *Vice: Dick Cheney and the Hijacking of the American Presidency* (New York: Random House, 2006). Ron Suskind, *The Price of Loyalty: George W. Bush, the White House, and the Education of Paul O'Neill* (New York: Simon & Schuster, 2004).

30. Jeffrey Kluger, "Bush's Hard Line Has Stunned Environmentalists, but with Concerted Action—and New Technologies—It's Not Too Late to Cool the Greenhouse," *Time*, April 9, 2001. "The Kyoto Protocol—Status of Ratification | UNFCCC," accessed July 13, 2020, https://unfccc.int/process/the-kyoto-protocol/status-of-ratification.

31. The emissions change of 23 percent is from R. Stavins et al., "International Cooperation: Agreements and Instruments," Chapter 13 in *Climate*

Change 2014: Mitigation of Climate Change. Contribution of Working Group III to the Fifth Assessment Report of the Intergovernmental Panel on Climate Change (Cambridge: Cambridge University Press, 2014), 1043. The 23 percent change includes land use and forestry; excluding those, the reduction was 17 percent.

Discussion of compliance and the "hot air" quote is from Igor Shishlov, Romain Morel, and Valentin Bellassen, "Compliance of the Parties to the Kyoto Protocol in the First Commitment Period," *Climate Policy* 16, no. 6 (August 17, 2016): 768–82, https://doi.org/10.1080/14693062.2016.1164658.

32. All data are from UN Framework Convention on Climate Change. The U.S. trend of 12 percent is based on total greenhouse gases including land use, land cover, and forestry; without them, growth was 9 percent: https://di.unfccc.int/time_series. China and Korea trends are from https://di.un fccc.int/ghg_profile_non_annex1. Global Carbon Project, "Supplemental Data of Global Carbon Budget 2018 (Version 1.0) [Data Set]," 2018, https://doi.org/10.18160/gcp-2018. Raphael Calel and Antoine Dechezleprêtre, "Environmental Policy and Directed Technological Change: Evidence from the European Carbon Market," *Review of Economics and Statistics* 98, no. 1 (June 24, 2014): 173–91, https://doi.org/10.1162/REST_a_00470.

33. Bodansky, "The Copenhagen Climate Change Conference."

34. Barack Obama, "President Obama Addresses Joint Session of Congress," February 4, 2009, www.washingtonpost.com/wp-srv/politics/documents/obama_address_022409.html. The G8 pledge: "We reiterate our willingness to share with all countries the goal of achieving at least a 50 percent reduction of global emissions by 2050. . . . We also support a goal of developed countries reducing emissions of greenhouse gases in aggregate by 80 percent or more by 2050 compared to 1990 or more recent years." www.g8.utoronto.ca/summit/2009laquila/2009-declaration.html.

35. "Obama's Speech to the Copenhagen Climate Summit | Environment | The Guardian," December 18, 2009, www.theguardian.com/environ ment/2009/dec/18/obama-speech-copenhagen-climate-summit. Massachusetts v. EPA, 549 U.S. 497 (2007).

36. Sweet, *Climate Diplomacy from Rio to Paris*, Chapter 9.

37. Sweet, *Climate Diplomacy from Rio to Paris*, 149.

38. Busby and Purvis, "Climate Leadership in Uncertain Times."

39. Sweet, *Climate Diplomacy from Rio to Paris*, 161.

40. Kimble, interview.

41. "U.S.-China Joint Announcement on Climate Change," whitehouse.gov, November 11, 2014, https://obamawhitehouse.archives.gov/the-press-office/2014/11/11/us-china-joint-announcement-climate-change.

42. Sweet, *Climate Diplomacy from Rio to Paris*, 172. Radoslav S. Dimitrov, "The Paris Agreement on Climate Change: Behind Closed Doors," *Global Environmental Politics* 16, no. 3 (July 15, 2016): 1–11, https://doi. org/10.1162/GLEP_a_00361.

43. Todd Stern is quoted in Fred Dews, Todd Stern, and David Wessel, "One Year After Trump's Decision to Leave the Paris Climate Agreement," Brookings Cafeteria, May 13, 2018, 13. Dimitrov, "The Paris Agreement on Climate Change."

44. Kimble, interview. Climate Action Tracker, "Warming Projections Global Update: Governments Still Showing Little Sign of Acting on Climate Crisis." United Nations Environment Programme, "Emissions Gap Report 2019" (Nairobi: UNEP, 2019), www.unenvironment.org/resources/ emissions-gap-report-2019.

45. "Paris Agreement Signatories," accessed July 13, 2020, https://treaties. un.org/Pages/ViewDetails.aspx?src=TREATY&mtdsg_no=XXVII-7- d&chapter=27&clang=_en.

46. William D. Nordhaus, "Can We Control Carbon Dioxide?," *IAASA Working Paper*, 1975, 49. Nordhaus used similar language in William D. Nordhaus, "Strategies for the Control of Carbon Dioxide," Cowles Foundation Discussion Paper (Cowles Foundation for Research in Economics, Yale University, 1977), https://econpapers.repec.org/paper/ cwlcwldpp/443.htm., which included a figure labeled with the 2C limit. He used similar language again in William D. Nordhaus, *The Efficient Use of Energy Resources* (New Haven: Yale University Press, 1980), 140–42. More recent estimates of temperature ranges are from Valérie Masson-Delmotte et al., "Information from Paleoclimate Archives," Chapter 5 in *Climate Change 2013: The Physical Science Basis. Contribution of Working Group I to the Fifth Assessment Report of the Intergovernmental Panel on Climate Change* (Cambridge: Cambridge University Press, 2013), 383–464.

47. Nordhaus, "Can We Control Carbon Dioxide?," 24. The criticism of warming limits is from David G. Victor and Charles F. Kennel, "Climate Policy: Ditch the 2 °C Warming Goal," *Nature News* 514, no. 7520 (October 2, 2014): 30, https://doi.org/10.1038/514030a.

48. Carlo C. Jaeger and Julia Jaeger, "Three Views of Two Degrees," *Regional Environmental Change* 11, no. S1 (March 2011): 15–26, https://doi. org/10.1007/s10113-010-0190-9. John Bachmann, "Will the Circle Be Unbroken: A History of the U.S. National Ambient Air Quality Standards," *Journal of the Air & Waste Management Association* 57, no. 6 (June 1, 2007): 652–97, https://doi.org/10.3155/1047-3289.57.6.652.

49. Robin Webster, "A Brief History of the 1.5C Target," *Climate Home News*, December 10, 2015, www.climatechangenews.com/2015/12/10/a-brief -history-of-the-1-5c-target. Timothée Ourbak and Alexandre K. Magnan, "The Paris Agreement and Climate Change Negotiations: Small Islands, Big Players," *Regional Environmental Change* 18, no. 8 (December 2018): 2201–7, https://doi.org/10.1007/s10113-017-1247-9. "Linguistic gymnastics" is from Dimitrov, "The Paris Agreement on Climate Change." NOAA National Centers for Environmental Information, Climate at a Glance: Global Time Series, published July 2020, retrieved on July 17, 2020, from www.ncdc.noaa.gov/cag. Mooney, Freedman, and Muyskens, "2020 Rivals Hottest Year on Record, Pushing Earth Closer to a Critical Climate Threshold." Rohde, "Global Temperature Report for 2020."

50. Oliver Geden, Skype interview, May 22, 2018.

51. J. Fuglestvedt et al., "Implications of Possible Interpretations of 'Greenhouse Gas Balance' in the Paris Agreement," *Philosophical Transactions of the Royal Society A: Mathematical, Physical and Engineering Sciences* 376, no. 2119 (May 13, 2018): 20160445, https://doi.org/10.1098/rsta.2016.0445.

52. IPCC, "Global Warming of 1.5°C." Levin et al., "Designing and Communicating Net-Zero Targets."

53. Daniel Bodansky, Jutta Brunnee, and Lavanya Rajamani, *International Climate Change Law* (Oxford: Oxford University Press, 2017). Fuglestvedt et al., "Implications of Possible Interpretations of 'Greenhouse Gas Balance' in the Paris Agreement." IPCC, *Climate Change 2013: The Physical Science Basis. Contribution of Working Group I to the Fifth Assessment Report of the Intergovernmental Panel on Climate Change*, ed. T. F. Stocker, D. Qin, and G.-K. Plattner (Cambridge: Cambridge University Press, 2013).

54. Kelly Levin, Email, December 4, 2020.

55. Parties initially submit an intended nationally determined contribution (INDC), which becomes a nationally determined contribution (NDC) after they formally enter the Paris Agreement, as explained at www.wri. org/indc-definition. For simplicity, I'll refer to pledges as NDCs or contributions. Temperature projections are from Climate Action Tracker, "Warming Projections Global Update: Governments Still Showing Little Sign of Acting on Climate Crisis." David Victor, Phone interview, November 16, 2018.

56. White House, "United States Mid-Century Strategy for Deep Decarbonization," November 2016, https://unfccc.int/files/focus/long-term_strate gies/application/pdf/mid_century_strategy_report-final_red.pdf. UN-FCCC, "NDC Registry," accessed July 14, 2020, www4.unfccc.int/sites/ ndcstaging/Pages/Home.aspx. "Fact Sheet: President Biden Sets 2030

Greenhouse Gas Pollution Reduction Target Aimed at Creating Good-Paying Union Jobs and Securing U.S. Leadership on Clean Energy Technologies," White House, April 22, 2021, www.whitehouse.gov/briefing-room/statements-releases/2021/04/22/fact-sheet-president-biden-sets-2030-greenhouse-gas-pollution-reduction-target-aimed-at-creating-good-paying-union-jobs-and-securing-u-s-leadership-on-clean-energy-technologies.

57. Environmental Protection Agency, "Inventory of U.S. Greenhouse Gas Emissions and Sinks: 1990–2019," April 2021, www.epa.gov/ghgemissions/inventory-us-greenhouse-gas-emissions-and-sinks. Energy Information Administration, Annual Energy Outlook 2021. Climate Action Tracker, "Warming Projections Global Update: Governments Still Showing Little Sign of Acting on Climate Crisis."

58. Susan Biniaz, "Broadening Action on Climate Change: The Paris Agreement Cannot Do It Alone," in *A Better Planet: Forty Ideas for a Sustainable Future*, ed. Daniel C. Esty (New Haven: Yale University Press, 2019), https://yalebooks.yale.edu/book/9780300246247/better-planet.

Chapter 3. The Road from Paris

1. Robert Axelrod, "Effective Choice in the Prisoner's Dilemma," *Journal of Conflict Resolution* 24, no. 1 (1980): 3–25. Robert Axelrod, "More Effective Choice in the Prisoner's Dilemma," *Journal of Conflict Resolution* 24, no. 3 (1980): 379–403. In an email dated June 20, 2019, Axelrod confirmed that it was paper invitations.

2. Each strategy was also paired with itself and a program that issued random moves.

3. Email from Robert Axelrod (June 20, 2019) confirmed that this was a surprise.

4. Robert Axelrod, *The Evolution of Cooperation* (New York: Basic Books, 1984).

5. This wording is from Michael Liebreich, "How to Save the Planet: Be Nice, Retaliatory, Forgiving & Clear," September 11, 2007, https://about.bnef.com/blog/how-to-save-the-planet-be-nice-retaliatory-forgiving-and-clear.

6. Aklin and Mildenberger, "Prisoners of the Wrong Dilemma." Michael Liebreich, "How to Save the Planet."

7. Aklin and Mildenberger, "Prisoners of the Wrong Dilemma." "The Peoples' Climate Vote" (United Nations Development Programme, 2021), www.undp.org/content/undp/en/home/librarypage/climate-and-disaster-

resilience-/The-Peoples-Climate-Vote-Results.html. Jonathon P. Schuldt and Y. Connie Yuan, "Despite What Trump Says, Most Americans Want Climate Action—Even If China Doesn't Do Its Part," *Washington Post*, January 13, 2019, www.washingtonpost.com/news/monkey-cage/wp/2019/01/03/despite-what-trump-says-most-americans-want-climate-action-even-if-china-doesnt-do-its-part. Danny Cullenward and David G. Victor, *Making Climate Policy Work* (Cambridge, UK: Polity Press, 2020).

8. Michael Liebreich, "How to Save the Planet."

9. Benjamin Sovacool, Phone interview, June 25, 2018.

10. Henry A. Waxman, "H.R. 2454—111th Congress (2009–2010): American Clean Energy and Security Act of 2009" (2009), www.congress.gov/bill/111th-congress/house-bill/2454. James A. Baker et al., "The Conservative Case for Carbon Dividends," February 2017, Climate Leadership Council, www.clcouncil.org/media/2017/03/The-Conservative-Case-for-Carbon-Dividends.pdf. Alexandria Ocasio-Cortez, "Text—H.Res.109— 116th Congress (2019–2020): Recognizing the Duty of the Federal Government to Create a Green New Deal" (2019), www.congress.gov/bill/116th-congress/house-resolution/109/text. Stephen Kho, Bernd G. Janzen, and Holly M. Smith, "Border Adjustment Measures in Proposed U.S. Climate Change Legislation—'A New Chapter in America's Leadership on Climate Change?,' " *Sustainable Development Law & Policy* (Spring 2009): 12–19, 59–62. Benjamin Sovacool, interview. Busby, "A Green Giant? Inconsistency and American Environmental Diplomacy."

11. Frédéric Simon, "French to Revive Sarkozy's EU Carbon Tariff Idea," *Www.Euractiv.Com*, May 18, 2012, sec. Climate change, www.euractiv.com/section/climate-environment/news/french-to-revive-sarkozy-s-eu-carbon-tariff-idea. Mia Shanley and Ilona Wissenbach, "Germany Calls Carbon Tariffs 'Eco-Imperialism,' " *Reuters*, July 24, 2009, www.reuters.com/article/us-germany-tariffs-idUSTRE56N1RJ20090724. Benjamin Kentish, "Nicolas Sarkozy Promises to Hit America with a Carbon Tax If Donald Trump Rips Up Landmark Paris Climate Deal," *The Independent*, November 15, 2016, sec. Europe, www.independent.co.uk/news/world/europe/donald-trump-us-carbon-tax-nicolas-sarkozy-global-warming-paris-climate-deal-a7418301.html.

12. European Commission, "The European Green Deal," December 11, 2019, https://eur-lex.europa.eu/legal-content/EN/TXT/PDF/?uri=CELEX:52019DC0640&from=EN.

13. Kate Abnett, "EU Sees Carbon Border Levy as 'Matter of Survival' for Industry," *Reuters*, January 18, 2021, www.reuters.com/article/us-climate-change-eu-carbon-idUSKBN29N1R1. Josh Gabbatiss, "Q&A: What

Does the Brexit Deal Say About Climate Change and Energy?," *Carbon Brief*, January 18, 2021, www.carbonbrief.org/qa-what-does-the-brexit-deal-say-about-climate-change-and-energy.

14. Sweet, *Climate Diplomacy from Rio to Paris*. Climate Action Tracker, "China Going Carbon Neutral Before 2060 Would Lower Warming Projections by Around 0.2 to 0.3 Degrees C," September 23, 2020, https://climateactiontracker.org/press/china-carbon-neutral-before-2060-would-lower-warming-projections-by-around-2-to-3-tenths-of-a-degree.

15. Aklin and Mildenberger, "Prisoners of the Wrong Dilemma."

16. Kimble, interview.

17. Simon Evans and Jocelyn Timperley, "COP24: Key Outcomes Agreed at the UN Climate Talks in Katowice," *Carbon Brief*, December 16, 2018, www.carbonbrief.org/cop24-key-outcomes-agreed-at-the-un-climate-talks-in-katowice. Simon Evans and Josh Gabbatiss, "COP25: Key Outcomes Agreed at the UN Climate Talks in Madrid," *Carbon Brief*, December 15, 2019, www.carbonbrief.org/cop25-key-outcomes-agreed-at-the-un-climate-talks-in-madrid.

18. Christiana Figueres, "Climate Change Is Speeding Up. Our Response Needs to Be Even Faster," World Economic Forum, September 7, 2018, www.weforum.org/agenda/2018/09/why-there-need-for-speed-on-climate-action-paris-agreement.

19. Parkash Chander, *Game Theory and Climate Change* (New York: Columbia University Press, 2018).

20. Simon Evans, Email, July 17, 2020.

21. Paris Agreement Articles 7, 9, and 10. Evans and Timperley, "COP24." Eliza Northrop et al., "Achieving the Ambition of Paris: Designing the Global Stocktake" (World Resources Institute, May 2018), https://wriorg.s3.amazonaws.com/s3fs-public/achieving-ambition-paris-designing-global-stockade.pdf.

22. IPCC, "Global Warming of 1.5°C." United Nations Environment Programme, "Emissions Gap Report 2020." Steven Lee Myers, "China's Pledge to Be Carbon Neutral by 2060: What It Means," *New York Times*, September 23, 2020, sec. World, www.nytimes.com/2020/09/23/world/asia/china-climate-change.html.

23. "U.S.-China Joint Announcement on Climate Change."

24. Victor, interview. UNFCCC, "NDC Registry." United Nations Environment Programme, "Emissions Gap Report 2020."

25. Biniaz, interview.

26. Aklin and Mildenberger, "Prisoners of the Wrong Dilemma."

27. Arild Underdal, *The Politics of International Fisheries Management: The Case of the Northeast Atlantic* (Universitetsforl., 1980). Sweet, *Climate Diplomacy from Rio to Paris*. Robinson Meyer, "A Reader's Guide to the Paris Agreement," *The Atlantic*, December 16, 2015, www.theatlantic.com/science/archive/2015/12/a-readers-guide-to-the-paris-agreement/420345.

28. Michael Taylor, "Norway Starts Payments to Indonesia for Cutting Forest Emissions," *Reuters*, February 18, 2019, www.reuters.com/article/us-indonesia-climatechange-forests-idUSKCN1Q70ZY. Shannon N. Koplitz et al., "Public Health Impacts of the Severe Haze in Equatorial Asia in September–October 2015: Demonstration of a New Framework for Informing Fire Management Strategies to Reduce Downwind Smoke Exposure," *Environmental Research Letters* 11, no. 9 (September 2016): 094023, https://doi.org/10.1088/1748-9326/11/9/094023.

29. William Nordhaus, "Climate Clubs: Overcoming Free-Riding in International Climate Policy," *American Economic Review* 105, no. 4 (April 2015): 1339–70, https://doi.org/10.1257/aer.15000001. William D. Nordhaus, "Climate Change: The Ultimate Challenge for Economics; Nobel Lecture in Economic Sciences" (Stockholm University, December 8, 2018), www.nobelprize.org/uploads/2018/10/nordhaus-slides.pdf.

30. Matthias Weitzel, Michael Hübler, and Sonja Peterson, "Fair, Optimal, or Detrimental? Environmental vs. Strategic Use of Border Carbon Adjustment," *Energy Economics*, The Role of Border Carbon Adjustment in Unilateral Climate Policy: Results from EMF 29, 34 (December 1, 2012): S198–207, https://doi.org/10.1016/j.eneco.2012.08.023. Christoph Böhringer, Jared C. Carbone, and Thomas F. Rutherford, "Unilateral Climate Policy Design: Efficiency and Equity Implications of Alternative Instruments to Reduce Carbon Leakage," *Energy Economics*, The Role of Border Carbon Adjustment in Unilateral Climate Policy: Results from EMF 29, 34 (December 1, 2012): S208–17, https://doi.org/10.1016/j.eneco.2012.09.011. Arvind Ravikumar, email interview, August 10, 2020.

31. Victor, *Global Warming Gridlock*. Victor, interview.

32. Gregory F. Nemet, Vera Zipperer, and Martina Kraus, "The Valley of Death, the Technology Pork Barrel, and Public Support for Large Demonstration Projects," *Energy Policy* 119 (August 2018): 154–67, https://doi.org/10.1016/j.enpol.2018.04.008.

33. Detlef F. Sprinz et al., "The Effectiveness of Climate Clubs Under Donald Trump," *Climate Policy* 18, no. 7 (August 9, 2018): 828–38, https://doi.org/10.1080/14693062.2017.1410090. Jon Hovi, Skype interview, August 20, 2019. "Trade and Cooperation Agreement Between the European Union and the European Atomic Energy Community of the One Part, and the

United Kingdom of Great Britain and Northern Ireland, of the Other Part" (Brussels: European Commission, 2020), https://ec.europa.eu/transparency/regdoc/rep/1/2020/EN/COM-2020-857-F1-EN-ANNEX-1-PART-1.PDF.

34. UN Environment Programme, "Kigali Amendment Hits Milestone 100th Ratification, Boosting Climate Action," July 14, 2020, www.unenvironment.org/news-and-stories/press-release/kigali-amendment-hits-milestone-100th-ratification-boosting-climate. Benjamin J. Hulac, "Trump Drags Feet on Climate Treaty, and Republicans Aren't Happy," *Roll Call*, May 13, 2019, www.rollcall.com/2019/05/13/trump-drags-feet-on-climate-treaty-and-republicans-arent-happy.

35. Julian Kirchherr and Frauke Urban, "Technology Transfer and Cooperation for Low Carbon Energy Technology: Analysing 30 Years of Scholarship and Proposing a Research Agenda," *Energy Policy* 119 (August 1, 2018): 600–609, https://doi.org/10.1016/j.enpol.2018.05.001. Jessica Green, Phone interview, August 27, 2019.

36. Kirchherr and Urban, "Technology Transfer and Cooperation for Low Carbon Energy Technology."

37. Geir Asheim, Phone interview, August 16, 2019.

38. Christophe McGlade and Paul Ekins, "The Geographical Distribution of Fossil Fuels Unused When Limiting Global Warming to 2 °C," *Nature* 517 (January 7, 2015): 187. SEI et al., "The Production Gap Report: 2020 Special Report," 2020, http://productiongap.org/2020report.

39. Hans-Werner Sinn, *The Green Paradox: A Supply-Side Approach to Global Warming* (Cambridge: MIT Press, 2012).

40. Christopher Ingraham, "Analysis | The Entire Coal Industry Employs Fewer People than Arby's," *Washington Post*, accessed July 14, 2020, www.washingtonpost.com/news/wonk/wp/2017/03/31/8-surprisingly-small-industries-that-employ-more-people-than-coal. S&P Global Market Intelligence, "U.S. Coal Companies Continue to Lose Market Value," *Institute for Energy Economics & Financial Analysis* (blog), January 11, 2019, https://ieefa.org/u-s-coal-companies-continue-to-lose-market-value. "Market Value of Coal Mining in the United States from 2010 to 2020," Statista, August 6, 2020, www.statista.com/statistics/1137311/market-size-of-coal-mining-in-the-us.

41. EIA, "U.S. Coal Reserves—U.S. Energy Information Administration (EIA)," October 3, 2019, www.eia.gov/coal/reserves. EIA, "Coal Explained: How Much Coal Is Left?," November 12, 2019, www.eia.gov/energyexplained/coal/how-much-coal-is-left.php.

42. Daniel Cohan, "The Big New Concept in Obama's Keystone XL Decision," *Houston Chronicle*, November 16, 2015, www.houstonchronicle.com/

local/gray-matters/article/Keystone-6625749.php. Georgia Piggot, Phone interview, August 27, 2019. White House, "Executive Order on Protecting Public Health and the Environment and Restoring Science to Tackle the Climate Crisis," January 20, 2021, www.whitehouse.gov/brief ing-room/presidential-actions/2021/01/20/executive-order-protecting-public-health-and-environment-and-restoring-science-to-tackle-cli-mate-crisis. Coral Davenport, Henry Foutain, and Lisa Friedman, "Biden Suspends Drilling Leases in Arctic National Wildlife Refuge," *New York Times*, June 1, 2021, www.nytimes.com/2021/06/01/climate/biden-drilling-arctic-national-wildlife-refuge.html.

43. Bård Harstad, "Buy Coal! A Case for Supply-Side Environmental Policy," *Journal of Political Economy* 120, no. 1 (February 1, 2012): 77–115, https://doi. org/10.1086/665405. Bård Harstad, "Making Paris Sustainable," in *The Paris Agreement and Beyond: International Climate Change Policy Post-2020*, ed. Robert N. Stavins and Robert C. Stowe (Cambridge: Harvard Project on Climate Agreements, 2016), 33–36. G. B. Asheim et al., "The Case for a Supply-Side Climate Treaty," *Science* 365, no. 6451 (July 26, 2019): 325–27, https://doi.org/10.1126/science.aax5011. Victor, interview. Asheim, interview.

Chapter 4. *Pillars of Decarbonization*

1. David Mikkelson, "Willie Sutton—'That's Where the Money Is,'" *Snopes*, November 14, 2008, www.snopes.com/fact-check/willie-sutton. Environmental Protection Agency, "Inventory of U.S. Greenhouse Gas Emissions and Sinks: 1990–2019." EPA tallies 6.6 billion metric tons of carbon dioxide equivalents, minus 0.8 billion tons of land-use related sinks. Olivier and Peters, "Trends in Global CO_2 and Total Greenhouse Gas Emissions."

2. EPA, "Inventory of U.S. Greenhouse Gas Emissions and Sinks: 1990–2019." "The Role of CCUS in Low-Carbon Power Systems" (Paris: International Energy Agency, July 2020), www.iea.org/reports/the-role-of-ccus-in-low-carbon-power-systems.

3. James H. Williams et al., "Carbon-Neutral Pathways for the United States," *AGU Advances* 2, no. 1 (2021): e2020AV000284, https://doi. org/10.1029/2020AV000284.

4. Adam B. Jaffe, Richard G. Newell, and Robert N. Stavins, "Environmental Policy and Technological Change," *Environmental and Resource Economics* 22, no. 1 (June 1, 2002): 41–70, https://doi.org/10.1023/A:1015519401088. Regarding the RD&D acronym, the International Energy Agency and

U.S. DOE define RD&D as "research, development, and demonstration," the California Public Utility Commission as "research, development, demonstration, and deployment," and the International Renewable Energy Agency as "research, development, and deployment."

5. Gregory F. Nemet, Vera Zipperer, and Martina Kraus, "The Valley of Death, the Technology Pork Barrel, and Public Support for Large Demonstration Projects," *Energy Policy* 119 (August 2018): 154–67, https://doi.org/10.1016/j.enpol.2018.04.008.

6. Nemet, Zipperer, and Kraus, "The Valley of Death, the Technology Pork Barrel, and Public Support for Large Demonstration Projects." Varun Sivaram et al., "To Bring Emissions-Slashing Technologies to Market, the United States Needs Targeted Demand-Pull Innovation Policies" (Columbia SIPA Center on Global Energy Policy, January 20, 2021).

7. Brian F. Gerke, "Light-Emitting Diode Lighting Products," in *Technological Learning in the Transition to a Low-Carbon Energy System: Conceptual Issues, Empirical Findings, and Use in Energy Modeling*, ed. M. Junginger (London: Academic Press Ltd–Elsevier Science Ltd, 2020), 233–56. Ambuj D. Sagar and Bob van der Zwaan, "Technological Innovation in the Energy Sector: R&D, Deployment, and Learning-by-Doing," *Energy Policy* 34, no. 17 (November 1, 2006): 2601–8, https://doi.org/10.1016/j.enpol.2005.04.012. Edward S. Rubin et al., "A Review of Learning Rates for Electricity Supply Technologies," *Energy Policy* 86 (November 2015): 198–218, https://doi.org/10.1016/j.enpol.2015.06.011. Nemet, Zipperer, and Kraus, "The Valley of Death, the Technology Pork Barrel, and Public Support for Large Demonstration Projects." Gregory F. Nemet, *How Solar Energy Became Cheap: A Model for Low-Carbon Innovation* (Routledge, 2019).

8. National Research Council, *Energy Research at DOE: Was It Worth It? Energy Efficiency and Fossil Energy Research, 1978 to 2000* (Washington, D.C.: National Academies Press, 2001), https://doi.org/10.17226/10165. NRC estimates $30 billion in economic benefits from $7 billion in expenses, with environmental benefits of $64–90 billion (pp. 63–64). IEA, "Global Reported Corporate Energy R&D Spending in Selected Sectors, 2010–2019," May 27, 2020, www.iea.org/data-and-statistics/charts/global-reported-corporate-energy-r-and-d-spending-in-selected-sectors-2010-2019. Colin Cunliff and Linh Nguyen, "Energizing Innovation: Raising the Ambition for Federal Energy RD&D in Fiscal Year 2022" (Information Technology & Innovation Foundation, May 2021), https://itif.org/publications/2021/05/17/energizing-innovation-raising-ambition-federal-energy-rdd-fiscal-year-2022. Varun Sivaram et al., *Energizing America: A Roadmap to Launch a*

National Energy Innovation Mission (New York: Columbia University SIPA Center on Global Energy Policy, 2020). National Academies of Sciences, Engineering, and Medicine, *Accelerating Decarbonization of the U.S. Energy System.* Sivaram et al., "To Bring Emissions-Slashing Technologies to Market, the United States Needs Targeted Demand-Pull Innovation Policies." "Energy Innovation: Developing the Technologies for Decarbonization" (American Energy Innovation Council, December 2020), http://ameri canenergyinnovation.org/2020/12/energy-innovation-developing-the-technologies-for-decarbonization.

9. Benjamin Gaddy, Varun Sivaram, and Francis O'Sullivan, "Venture Capital and Cleantech: The Wrong Model for Clean Energy Innovation" (MIT Energy Initiative, July 2016), https://energy.mit.edu/wp-content/uploads/2016/07/MITEI-WP-2016-06.pdf. Stephen Lacey, "Cleantech Venture Capital Is Back," *GreenTech Media*, February 12, 2019, www.greentechmedia.com/articles/read/cleantech-venture-capital-is-back. Jesse Jenkins and Sara Mansur, "Bridging the Clean Energy Valleys of Death: Helping American Entrepreneurs Meet the Nation's Energy Innovation Imperative" (Breakthrough Institute, 2011), https://thebreakthrough.org/articles/bridging-the-clean-energy-vall.

10. Amory Lovins, On-site visit, Basalt, Colorado, August 10, 2018. Joby Warrick, "Bill Gates: 'We Need an Energy Miracle' to Prevent Catastrophic Climate Change," *Washington Post*, February 23, 2016, www.washingtonpost.com/news/energy-environment/wp/2016/02/23/bill-gates-we-need-a-energy-miracle-to-prevent-catastrophic-climate-change.

11. Sarah Kaplan, "Biden Wants an All-Electric Federal Fleet. The Question Is: How Will He Achieve It?," *Washington Post*, January 28, 2021, www.washingtonpost.com/climate-solutions/2021/01/28/biden-federal-fleet-electric.

12. Meredith Fowlie, Michael Greenstone, and Catherine Wolfram, "Do Energy Efficiency Investments Deliver? Evidence from the Weatherization Assistance Program," *Quarterly Journal of Economics* 133, no. 3 (August 1, 2018): 1597–1644, https://doi.org/10.1093/qje/qjy005. Matt Viser and Dino Grandoni, "Biden, in New Climate Plan, Embraces More Aggressive Steps," *Washington Post*, July 14, 2020. Meredith Fowlie, "The Search for Good Green Stimulus," *Energy Institute Blog*, June 1, 2020, https://energyathaas.wordpress.com/2020/06/01/the-search-for-good-green-stimulus.

13. Climate Leadership Council, "Economists' Statement on Carbon Dividends," *Wall Street Journal*, January 17, 2019, https://clcouncil.org/economists-statement. "Energy Efficiency and Climate Change Mitigation: EMF Report 25, Volume I" (Stanford, Calif.: Energy Modeling Forum, March 2011). Rong Wang et al., "Induced Energy-Saving Efficiency

Improvements Amplify Effectiveness of Climate Change Mitigation," *Joule* 3, no. 9 (September 18, 2019): 2103–19, https://doi.org/10.1016/j.joule.2019.07.024.

14. Ben Evans, Phone interview, January 6, 2020.

15. According to EIA's "Monthly Energy Review," energy use in 2018 was just 1 percent higher than in 2005. EIA's Annual Energy Outlook 2020 projects energy use will increase 10 percent from 2019 to 2050. EIA has tended to underpredict improvements in technology; see Daniel Cohan, "The Trouble with Underestimating Clean Energy," *The Hill*, March 16, 2017, https://thehill.com/blogs/pundits-blog/energy-environment/322442-the-trouble-with-underestimating-clean-energy.

16. Steven Nadel and Lowell Ungar, "Halfway There: Energy Efficiency Can Cut Energy Use and Greenhouse Gas Emissions in Half by 2050" (American Council for an Energy-Efficient Economy, 2019), https://aceee.org/research-report/u1907. Amory Lovins, Marvin Odum, and John W. Rowe, *Reinventing Fire: Bold Business Solutions for the New Energy Era* (White River Junction, Vt.: Chelsea Green, 2013).

17. Steven Nadel, Phone interview, January 31, 2020.

18. Alan Kevin Meier, "Supply Curves of Conserved Energy" (Lawrence Berkeley Laboratory, University of California, 1982), https://escholarship.org/uc/item/20b1j10d. Alan Meier, Arthur H. Rosenfeld, and Janice Wright, "Supply Curves of Conserved Energy for California's Residential Sector," *Energy* 7, no. 4 (April 1, 1982): 347–58, https://doi.org/10.1016/0360-5442(82)90094-9. Hannah Choi Granade et al., "Unlocking Energy Efficiency in the U.S. Economy" (McKinsey & Company, July 2009). "Carbonomics Innovation, Deflation, and Affordable De-Carbonization" (Goldman Sachs, 2020), www.goldmansachs.com/insights/pages/gs-research/carbonomics-innovation-deflation-and-affordable-de-carbonization/report.pdf.

19. Amory B. Lovins, "How Big Is the Energy Efficiency Resource?," *Environmental Research Letters* 13, no. 9 (September 2018): 090401, https://doi.org/10.1088/1748-9326/aad965.

20. Amory Lovins, On-site visit, Basalt, Colorado, August 10, 2018. Lovins, "How Big Is the Energy Efficiency Resource?" Amory B. Lovins, "Reframing Automotive Fuel Efficiency," *SAE International Journal of Sustainable Transportation, Energy, Environment, & Policy* 1, no. 1 (April 16, 2020): 59–84, https://doi.org/10.4271/13-01-01-0004.

21. Hunt Allcott and Michael Greenstone, "Is There an Energy Efficiency Gap?," *Journal of Economic Perspectives* 26, no. 1 (February 2012): 3–28, https://doi.org/10.1257/jep.26.1.3. Adam B. Jaffe and Robert N. Stavins,

"The Energy-Efficiency Gap: What Does It Mean?," *Energy Policy* 22, no. 10 (October 1994): 804–10. Soren T. Anderson and Richard G. Newell, "Information Programs for Technology Adoption: The Case of Energy-Efficiency Audits" (Resources for the Future, September 2002), https://linkinghub.elsevier.com/retrieve/pii/S0928765503000484.

Chrishelle Lawrence, Maggie Woodward, and Chip Berry, "One in Eight U.S. Homes Uses a Programmed Thermostat with a Central Air Conditioning Unit—Today in Energy" (U.S. Energy Information Administration, July 19, 2017), www.eia.gov/todayinenergy/detail.php?id=32112.

22. Lorna A. Greening, David L. Greene, and Carmen Difiglio, "Energy Efficiency and Consumption—the Rebound Effect—a Survey," *Energy Policy* 28, no. 6 (June 1, 2000): 389–401, https://doi.org/10.1016/S0301-4215(00)00021-5. Steve Sorrell and John Dimitropoulos, "The Rebound Effect: Microeconomic Definitions, Limitations, and Extensions," *Ecological Economics* 65, no. 3 (April 15, 2008): 636–49, https://doi.org/10.1016/j.ecolecon.2007.08.013. Brinda A. Thomas and Inês L. Azevedo, "Estimating Direct and Indirect Rebound Effects for U.S. Households with Input-Output Analysis Part 1: Theoretical Framework," *Ecological Economics*, Sustainable Urbanisation: A Resilient Future, 86 (February 1, 2013): 199–210, https://doi.org/10.1016/j.ecolecon.2012.12.003. Steven Nadel, "The Rebound Effect: Large or Small?" (American Council for an Energy-Efficient Economy, August 2012), www.aceee.org/sites/default/files/pdf/white-paper/rebound-large-and-small.pdf. Greening, Greene, and Difiglio, "Energy Efficiency and Consumption—the Rebound Effect—a Survey." Kenneth Gillingham et al., "The Rebound Effect Is Overplayed," *Nature* 493, no. 7433 (January 2013): 475–76, https://doi.org/10.1038/493475a.

23. Bachmann, "Will the Circle Be Unbroken." Kateri Callahan, Phone interview, January 23, 2020. Environmental and Energy Study Institute, "Fact Sheet—Energy Efficiency Standards for Appliances, Lighting and Equipment," August 11, 2017, www.eesi.org/papers/view/fact-sheet-energy-efficiency-standards-for-appliances-lighting-and-equipment. Andrew DeLaski et al., "Next Generation Standards: How the National Energy Efficiency Standards Program Can Continue to Drive Energy, Economic, and Environmental Benefits" (American Council for an Energy-Efficient Economy, August 2016).

24. Department of Energy, "Energy Conservation Program: Energy Conservation Standards for General Service Lamps," *Federal Register* 84 FR 3120, no. 2019-01853 (February 11, 2019): 3120–31. U.S. Department of Energy, "Department of Energy Issues Final 'Process Rule' Modernizing Procedures in

the Consideration of Energy Conservation Standards," Energy.gov, January 15, 2020, www.energy.gov/articles/department-energy-issues-final-process-rule-modernizing-procedures-consideration-energy. Rebecca Tan, "Trump Blamed Energy-Saving Bulbs for Making Him Look Orange. Experts Say Probably Not," *Washington Post*, September 13, 2019, www.washingtonpost.com/politics/2019/09/13/trump-blamed-energy-saving-bulbs-making-him-look-orange-experts-say-probably-not. Hiroko Tabuchi, "Inside Conservative Groups' Effort to 'Make Dishwashers Great Again,' " *New York Times*, September 17, 2019, sec. Climate, www.nytimes.com/2019/09/17/climate/trump-dishwasher-regulatory-rollback.html. "Energy Efficiency Impact Report" (American Council for an Energy-Efficient Economy, Alliance to Save Energy, and The Business Council for Sustainable Energy, 2020), https://energyefficiencyimpact.org. Rebecca Beitsch, "Trump Administration Rolls Back Efficiency Standards for Showerheads, Washers, and Dryers," *The Hill*, December 15, 2020, https://thehill.com/policy/energy-environment/530310-trump-administration-rolls-back-efficiency-standards-for.

25. The International Council on Clean Transportation, "Chart Library: Passenger Vehicle Fuel Economy," 2020, https://theicct.org/chart-library-passenger-vehicle-fuel-economy. International Energy Agency, "Fuel Economy in Major Car Markets: Technology and Policy Drivers 2005–2017," March 2019, www.iea.org/topics/transport/gfei. Sandra Wappelhorst, "The End of the Road? An Overview of Combustion Engine Car Phase-out Announcements across Europe" (International Council on Clean Transportation, May 2020), https://theicct.org/sites/default/files/publications/Combustion-engine-phase-out-briefing-may11.2020.pdf. Anna Phillips and Russ Mitchell, "Trump Weakens Fuel Economy Standards, Rolling Back Key U.S. Effort Against Climate Change," *Los Angeles Times*, March 31, 2020, www.latimes.com/politics/story/2020-03-31/trump-rolls-back-fuel-economy-standards. EPA, "Inventory of U.S. Greenhouse Gas Emissions and Sinks: 1990–2019."

26. Environmental Protection Agency, "Final Rule for Phase 1 Greenhouse Gas Emissions Standards and Fuel Efficiency Standards for Medium- and Heavy-Duty Engines and Vehicles" (2011). Environmental Protection Agency, "Final Rule for Phase 2 Greenhouse Gas Emissions Standards and Fuel Efficiency Standards for Medium-and Heavy-Duty Engines and Vehicles" (2016). Tim Dallmann and Jin Lingzhi, "Fuel Efficiency and Climate Impacts of Soot-Free Heavy-Duty Diesel Engines" (International Council on Clean Transportation, June 2020).

27. Umair Irfan, "Trump's Fight with California over Vehicle Emissions Rules Has Divided Automakers," *Vox*, November 5, 2019, www.vox.com/

policy-and-politics/2019/11/5/20942457/california-trump-fuel-economy-auto-industry. Hiroko Tabuchi, "New Rule in California Will Require Zero-Emissions Trucks," *New York Times*, June 25, 2020, sec. Climate, www.nytimes.com/2020/06/25/climate/zero-emissions-trucks-california.html. Sean O'Kane, "15 States Will Follow California's Push to Electrify Trucks and Buses," *The Verge*, July 14, 2020, www.theverge.com/2020/7/14/21324552/electric-trucks-buses-clean-air-zero-emissions-states.

28. Environmental Protection Agency, "EPA Proposes First Greenhouse Gas Emissions Standards for Aircraft," July 22, 2020, www.epa.gov/newsreleases/epa-proposes-first-greenhouse-gas-emissions-standards-aircraft.

29. Energy-Efficient Codes Coalition, "International Energy Conservation Code," n.d., https://energyefficientcodes.org/iecc. U.S. Department of Energy, "Status of Energy Code Adoption," accessed July 24, 2020, www.energycodes.gov/status-state-energy-code-adoption. California Energy Commission, "California Building Energy Efficiency Standards—Title 24," accessed July 24, 2020, www.energy.ca.gov/programs-and-topics/programs/building-energy-efficiency-standards. California Energy Commission Efficiency Division, "Frequently Asked Questions 2019 Building Energy Efficiency Standards," accessed June 2, 2021, www.energy.ca.gov/sites/default/files/2020-06/Title24_2019_Standards_detailed_faq_ada.pdf. Energy and Environmental Economics Inc., "Building Energy Efficiency Measure Proposal to the California Energy Commission for the 2019 Update to the Title 24 Part 6 Building Energy Efficiency Standards Rooftop Solar PV System," September 2017, https://efiling.energy.ca.gov/GetDocument.aspx?tn=222201&DocumentContentId=27371. California Energy Commission Efficiency Division and California Public Utilities Commission Energy Division, "New Residential Zero Net Energy Action Plan 2015–2020," June 2015, www.cpuc.ca.gov/uploadedFiles/CPUC_Public_Website/Content/Utilities_and_Industries/Energy/Energy_Programs/Demand_Side_Management/EE_and_Energy_Savings_Assist/ZNERESACTIONPLAN_FINAL_060815.pdf.

30. Georges Zissis and Paolo Bertoldi, "Status of LED—Lighting World Market in 2017" (European Commission, 2018). "Global Induction Cooktop Market Report 2020," March 30, 2020, www.researchreportsworld.com/-global-induction-cooktop-market-15503950. Energy Information Administration, "Annual Passenger Travel Tends to Increase with Income—Today in Energy," May 11, 2016, www.eia.gov/todayinenergy/detail.php?id=26192. Ben Evans, interview, January 6, 2020. Tabuchi, "Inside Conservative Groups' Effort to 'Make Dishwashers Great Again.'"

31. On-site visit to NREL, August 7, 2018. The World Bank, "Energy Intensity Level of Primary Energy (MJ/$2011 PPP GDP)—United States | Data," accessed July 24, 2020, https://data.worldbank.org/indicator/EG. EGY.PRIM.PP.KD?contextual=aggregate&locations=US.

32. The 100-year global warming potential of methane relative to carbon dioxide is 34–36, according to G. Myhre, D. Shindell, and F.-M. Breon, "Anthropogenic and Natural Radiative Forcing," in *Climate Change 2013: The Physical Science Basis. Contribution of Working Group I to the Fifth Assessment Report of the Intergovernmental Panel on Climate Change*, ed. T. F. Stocker and D. Qin (Cambridge: Cambridge University Press, 2013), 659–740.

33. Myhre, Shindell, and Breon, "Anthropogenic and Natural Radiative Forcing." EPA uses 25 as the 100-year global warming potential of methane, based on IPCC's 2007 assessment. Environmental Protection Agency, "Inventory of U.S. Greenhouse Gas Emissions and Sinks: 1990–2019." Ramón A. Alvarez et al., "Assessment of Methane Emissions from the U.S. Oil and Gas Supply Chain," *Science* 361, no. 6398 (July 13, 2018): 186–88, https://doi.org/10.1126/science.aar7204.

34. Environmental Protection Agency, "Inventory of U.S. Greenhouse Gas Emissions and Sinks: 1990-2019." Olivier and Peters, "Trends in Global CO2 and Total Greenhouse Gas Emissions." Alvarez et al., "Assessment of Methane Emissions from the U.S. Oil and Gas Supply Chain." Chelsea Harvey, "Methane Emissions from Oil and Gas May Be Significantly Underestimated," *Scientific American*, February 21, 2020, www.scientificamerican.com/article/methane-emissions-from-oil-and-gas-may-be-significantly-underestimated. Samantha Mathewson, "MethaneSAT Picks SpaceX for Satellite Launch to Track Methane Levels in Earth's Atmosphere," *Space.Com*, January 13, 2021, www.space.com/methanesat-picks-spacex-methane-satellite-launch-2022.

35. Sarah White and Scott DiSavino, "France Halts Engie's U.S. LNG Deal amid Trade, Environment Disputes," *Reuters*, October 22, 2020, www.reuters.com/article/engie-lng-france-unitedstates-idUSKBN27808G.

36. Tapan K. Adhya et al., "Wetting and Drying: Reducing Greenhouse Gas Emissions and Saving Water from Rice Production" (World Resources Institute, December 2014). "USDA Building Blocks for Climate Smart Agriculture and Forestry Implementation Plan and Progress Report" (U.S. Department of Agriculture, May 2016). Robert D. Kinley et al., "Mitigating the Carbon Footprint and Improving Productivity of Ruminant Livestock Agriculture Using a Red Seaweed," *Journal of Cleaner Production* 259 (June 20, 2020): 120836, https://doi.org/10.1016/j.jclepro.2020.120836.

37. Peter Tschofen, Inês L. Azevedo, and Nicholas Z. Muller, "Fine Particulate Matter Damages and Value Added in the U.S. Economy," *Proceedings of the National Academy of Sciences* 116, no. 40 (October 1, 2019): 19857, https://doi.org/10.1073/pnas.1905030116. "USDA Building Blocks for Climate Smart Agriculture and Forestry Implementation Plan and Progress Report." Xiang Yan et al., "Recent Advances on the Technologies to Increase Fertilizer Use Efficiency," *Agricultural Sciences in China* 7, no. 4 (April 1, 2008): 469–79, https://doi.org/10.1016/S1671-2927(08)60091-7.

38. Jaclyn Kahn, "Renewable Natural Gas: Up-and-Coming Renewable Energy Contender?," *National Conference of State Legislatures Blog*, 2020, www.ncsl.org/blog/2020/08/10/renewable-natural-gas-up-and-coming-renewable-energy-contender.aspx.

39. Bingli Clark Chai et al., "Which Diet Has the Least Environmental Impact on Our Planet? A Systematic Review of Vegan, Vegetarian, and Omnivorous Diets," *Sustainability* 11, no. 15 (January 2019): 4110, https://doi.org/10.3390/su11154110. U.S. Department of Agriculture, "Food Waste FAQs," accessed July 24, 2020, www.usda.gov/foodwaste/faqs.

40. UN Environment Programme, "Kigali Amendment Hits Milestone 100th Ratification, Boosting Climate Action." Robert Beverly, "Congress Approves HFC Phasedown Plan in Omnibus Bill," *Air Conditioning Heating Refrigeration News*, December 22, 2020, www.achrnews.com/articles/144236-congress-approves-hfc-phasedown-plan-in-omnibus-bill?v=preview.

Chapter 5. Decarbonizing Electricity

1. White House, "United States Mid-Century Strategy for Deep Decarbonization." Deep Decarbonization Pathways Project, "Pathways to Deep Decarbonization 2015 Report—Executive Summary" (SDSN—IDDRI, 2015). Larson et al., "Net-Zero America." Steven J. Davis et al., "Net-Zero Emissions Energy Systems," *Science* 360, no. 6396 (June 29, 2018), https://doi.org/10.1126/science.aas9793. Ben Haley et al., "350 PPM Pathways for the United States" (Deep Decarbonization Pathways Project, May 8, 2019). The Zero Carbon Consortium, "America's Zero Carbon Action Plan."

2. Energy Information Administration, "Monthly Energy Review."

3. Peter Behr, Edward Klump, and Lesley Clark, "Politics: Is Biden's 100% Clean Electricity Plan Doable?," *E&E News*, July 15, 2020, www.eenews.net/stories/1063565769. Viser and Grandoni, "Biden, in New Climate Plan, Embraces More Aggressive Steps." Ocasio-Cortez, "Recognizing

the Duty of the Federal Government to Create a Green New Deal." National Conference of State Legislatures, "State Renewable Portfolio Standards and Goals," April 17, 2020, www.ncsl.org/research/energy/renewable-portfolio-standards.aspx. Joe Biden, "Plan for Climate Change and Environmental Justice," Joe Biden for President: Official Campaign Website, 2020, https://joebiden.com/climate-plan.

4. Energy Information Administration, "Monthly Energy Review." Jim Giles, "Methane Quashes Green Credentials of Hydropower," *Nature* 444, no. 7119 (November 30, 2006): 524–25, https://doi.org/10.1038/444524a. Ilissa B. Ocko and Steven P. Hamburg, "Climate Impacts of Hydropower: Enormous Differences Among Facilities and over Time," *Environmental Science & Technology* 53, no. 23 (December 3, 2019): 14070–82, https://doi.org/10.1021/acs.est.9b05083.

5. U.S. Department of Energy, "Hydropower Vision," 2016, www.energy.gov/sites/prod/files/2018/02/f49/Hydropower-Vision-021518.pdf.

6. "Levelized Cost and Levelized Avoided Cost of New Generation Resources" (Energy Information Administration, February 2020), www.eia.gov/outlooks/aeo/pdf/electricity_generation.pdf. Douglas Hall, Kelly Reeves, and Julie Brizzee, "Feasibility Assessment of the Water Energy Resources of the United States for New Low Power and Small Hydro Classes of Hydroelectric Plants" (U.S. Department of Energy, January 2006), www.energy.gov/sites/prod/files/2014/05/f15/doewater-11263.pdf. Shih-Chieh Kao et al., "New Stream-Reach Development: A Comprehensive Assessment of Hydropower Energy Potential in the United States," April 1, 2014, https://doi.org/10.2172/1130425.

7. Thomas Wellock, " 'Too Cheap to Meter': A History of the Phrase," *U.S. NRC Blog* (blog), June 3, 2016, https://public-blog.nrc-gateway.gov/2016/06/03/too-cheap-to-meter-a-history-of-the-phrase. National Renewable Energy Laboratory, "Life Cycle Assessment Harmonization," accessed July 26, 2020, www.nrel.gov/analysis/life-cycle-assessment.html. Nuclear capacity is from EIA's "Monthly Energy Review," and projections are from Paul L. Joskow and Martin L. Baughman, "The Future of the U.S. Nuclear Energy Industry," *Bell Journal of Economics* 7, no. 1 (1976): 3–32, https://doi.org/10.2307/3003188. Even before Three Mile Island, in 1978, projections had already been scaled back by more than half, to about 300 GW by 2000, due to concerns about cost, construction delays, and lack of long-term waste disposal, according to EIA Annual Report to Congress 1978, 3:223.

8. Mycle Schneider and Antony Froggatt, "The World Nuclear Industry Status Report 2019," September 27, 2019, www.worldnuclearreport.org/

The-World-Nuclear-Industry-Status-Report-2019-html. World Nuclear Association, "Plans for New Nuclear Reactors Worldwide," May 2021, www.world-nuclear.org/information-library/current-and-future-genera tion/plans-for-new-reactors-worldwide.aspx.

9. United States Nuclear Regulatory Commission, "Backgrounder on Reactor License Renewal," October 1, 2018, www.nrc.gov/reading-rm/ doc-collections/fact-sheets/fs-reactor-license-renewal.html. World Nuclear Association, "Nuclear Power in the USA," May 2020, www. world-nuclear.org/information-library/country-profiles/countries-t-z/ usa-nuclear-power.aspx. Energy Information Administration, "Monthly Energy Review."

10. Energy Information Administration, "Monthly Energy Review." Energy Information Administration, "America's Oldest Operating Nuclear Power Plant to Retire on Monday—Today in Energy," September 14, 2018, www. eia.gov/todayinenergy/detail.php?id=37055. Energy Information Administration, "Preliminary Monthly Electric Generator Inventory: Form EIA-860M," March 2021, www.eia.gov/electricity/data/eia860m. Mark Chediak and Gerson Freitas Jr., "Bribery Scandals Taint Efforts to Save U.S. Nuclear Plants," *BloombergQuint*, July 24, 2020, www.bloom bergquint.com/business/bribery-scandals-taint-efforts-to-save-u-s-nu clear-plants.

11. Brad Plumer, "U.S. Nuclear Comeback Stalls as Two Reactors Are Abandoned," *New York Times*, July 31, 2017, sec. Climate, www.nytimes. com/2017/07/31/climate/nuclear-power-project-canceled-in-south-caro lina.html. Andrew Brown and Avery G. Wilks, "Former SCANA Executive Pleads Guilty to Fraud Charges Tied to Failed SC Nuclear Project," *Post and Courier*, July 23, 2020, www.postandcourier.com/business/former-scana-executive-pleads-guilty-to-fraud-charges-tied-to-failed-sc-nuclear-project/article_26e23ca8-c50b-11ea-8377-e7b39854212b.html. Darrell Proctor, "Georgia PSC Backs Additional Costs for Vogtle Nuclear Project," *POWER*, February 19, 2019, www.powermag.com/georgia-psc-backs-additional-costs-for-vogtle-nuclear-project.

12. Matthew Yglesias, "An Expert's Case for Nuclear Power," *Vox*, February 28, 2020, www.vox.com/2020/2/28/21155995/jessica-lovering-nuclear-energy.

13. Sonal Patel, "NuScale, UAMPS Kick Off Idaho SMR Nuclear Plant Licensing," *POWER*, January 12, 2021, www.powermag.com/nuscale-uamps-kick-off-idaho-smr-nuclear-plant-licensing. "TerraPower, Wyoming Governor and PacifiCorp Announce Efforts to Advance Nuclear Technology in Wyoming," TerraPower, June 2, 2021, www.terrapower.com/

natrium-demo-wyoming-coal-plant. "Platts Insight: Nuclear Industry, Vendors Believe There's a Future for Microreactors," *S&P Platts Global Insight*, September 19, 2018, https://blogs.platts.com/2018/09/19/insight-nuclear-industry-vendors-microreactors.

14. EIA, "Levelized Cost and Levelized Avoided Cost of New Generation Resources."

15. Lois Parshley, "When It Comes to Nuclear Power, Could Smaller Be Better?," *Yale Environment 360*, February 19, 2020, https://e360.yale.edu/features/when-it-comes-to-nuclear-power-could-smaller-be-better. Lyman, quoted in Adrian Cho, "Smaller, Safer, Cheaper: One Company Aims to Reinvent the Nuclear Reactor and Save a Warming Planet," *Science*, February 21, 2019, www.sciencemag.org/news/2019/02/smaller-safer-cheaper-one-company-aims-reinvent-nuclear-reactor-and-save-warming-planet. Government Accountability Office, "Science and Tech Spotlight: Nuclear Microreactors," February 2020, www.gao.gov/assets/710/704824.pdf.

16. Alex Gilbert, Twitter direct messages, February 28, 2020.

17. Nuclear Energy Institute, "Nuclear Waste," accessed July 26, 2020, www.nei.org/fundamentals/nuclear-waste. Per Peterson, "Will the United States Need a Second Geologic Repository?," *The Bridge* 33, no. 3 (Fall 2003), www.nae.edu/7602/WilltheUnitedStatesNeedaSecondGeologic Repository. Michael Ford, Email, February 23, 2020.

18. Table 4-12 in IPCC, *Emissions Scenarios: Summary for Policymakers: A Special Report of IPCC Working Group III* (Geneva: World Meteorological Organization): UNEP (United Nations Environment Programme), 2000. Detlef P. van Vuuren et al., "The Representative Concentration Pathways: An Overview," *Climatic Change* 109, no. 1–2 (November 2011): 5–31, doi.org/10.1007/s10584-011-0148-z.

19. Gary T. Rochelle, "Amine Scrubbing for CO2 Capture," *Science* 325, no. 5948 (September 25, 2009): 1652–54, doi.org/10.1126/science.1176731.

20. NRG Energy, "Petra Nova: Carbon Capture and the Future of Coal Power," NRG Energy, accessed July 26, 2020, www.nrg.com/case-studies/petra-nova.html. Energy Information Administration, "Petra Nova Is One of Two Carbon Capture and Sequestration Power Plants in the World," October 31, 2017, www.eia.gov/todayinenergy/detail.php?id=33552. Jeremy Dillon and Carlos Anchondo, "Carbon Capture: Low Oil Prices Force Petra Nova into 'Mothball Status,'" *E&E News*, July 28, 2020, www.eenews.net/stories/1063645835. Dennis Wamstead and David Schlissel, "Petra Nova Mothballing Post-Mortem: Closure of Texas Carbon Capture Plant Is a Warning Sign" (Institute for Energy Economics and Financial Analysis, August 2020), https://ieefa.org/

wp-content/uploads/2020/08/Petra-Nova-Mothballing-Post-Mortem_
August-2020.pdf. Brian Strasert, Su Chen Teh, and Daniel S. Cohan, "Air
Quality and Health Benefits from Potential Coal Power Plant Closures in
Texas," *Journal of the Air & Waste Management Association* 69, no. 3 (March
4, 2019): 333–50, https://doi.org/10.1080/10962247.2018.1537984.

21. Meili Ding et al., "Carbon Capture and Conversion Using Metal—
Organic Frameworks and MOF-Based Materials," *Chemical Society Reviews* 48, no. 10 (May 20, 2019): 2783–2828, https://doi.org/10.1039/
C8CS00829A.

22. Katie Fehrenbacher, "Carbon Capture Suffers a Huge Setback as Kemper
Plant Suspends Work," *GTM*, June 29, 2017, www.greentechmedia.com/
articles/read/carbon-capture-suffers-a-huge-setback-as-kemper-plant-
suspends-work.

23. "FutureGen Fact Sheet: Carbon Dioxide Capture and Storage Project,"
MIT Carbon Capture and Sequestration Technologies, accessed July 26,
2020, https://sequestration.mit.edu/tools/projects/futuregen.html. White
House, "Interagency Carbon Capture and Storage Task Force," accessed
July 26, 2020, https://obamawhitehouse.archives.gov/node/10514.

24. Akshat Rathi, "U.S. Startup Plans to Build First Zero-Emission Gas
Power Plants," *Bloomberg*, April 15, 2021, www.bloomberg.com/news/
articles/2021-04-15/u-s-startup-plans-to-build-first-zero-emission-gas-
power-plants.

25. Lazard, "Levelized Cost of Energy and Levelized Cost of Storage—2020,"
October 19, 2020, www.lazard.com/perspective/levelized-cost-of-energy-
and-levelized-cost-of-storage-2020. Simon Evans and Josh Gabbatiss,
"Solar Is Now 'Cheapest Electricity in History,' Confirms IEA," *Carbon
Brief*, October 13, 2020, www.carbonbrief.org/solar-is-now-cheapest-elec
tricity-in-history-confirms-iea.

26. Ryan Wiser et al., "2018 Wind Technologies Market Report" (U.S. Department of Energy, Office of Energy Efficiency and Renewable Energy,
2019).

27. Lazard, "Levelized Cost of Energy and Levelized Cost of Storage—2020."

28. Nemet, *How Solar Energy Became Cheap.*

29. Nemet, *How Solar Energy Became Cheap.*

30. Nemet, *How Solar Energy Became Cheap.* Lazard, "Levelized Cost of Energy and Levelized Cost of Storage—2020."

31. David Moore, on-site interview, NREL, August 7, 2018. Varun Sivaram,
Taming the Sun: Innovations to Harness Solar Energy and Power the Planet
(Cambridge: MIT Press, 2018), https://mitpress.mit.edu/books/taming-
sun. Ian Maxwell, "Rule Number One: Never Bet Against Silicon," *Chem-*

istry in Australia, September 2014, 36–37. *How to Make Transportation Carbon Neutral?—David Keith,* 2019, www.youtube.com/watch?v =sYopOt9siLg.

32. Paul Denholm and Robert M. Margolis, "Land-Use Requirements and the per-Capita Solar Footprint for Photovoltaic Generation in the United States," *Energy Policy* 36, no. 9 (September 1, 2008): 3531–43, https://doi. org/10.1016/j.enpol.2008.05.035. Sivaram, interview. Larson et al., "Net-Zero America." Haley et al., "350 PPM Pathways for the United States." Mark Z. Jacobson et al., "Low-Cost Solution to the Grid Reliability Problem with 100% Penetration of Intermittent Wind, Water, and Solar for All Purposes," *Proceedings of the National Academy of Sciences* 112, no. 49 (December 8, 2015): 15060–65, https://doi.org/10.1073/pnas.1510028112. The Zero Carbon Consortium, "America's Zero Carbon Action Plan." EIA Electric Power Monthly data are for twelve months ending October 2020 and include small-scale solar.

33. Sivaram, interview. Moore, interview. Chase, quoted in Andy Extance, "The Reality Behind Solar Power's Next Star Material," *Nature* 570, no. 7762 (June 25, 2019): 429–32, https://doi.org/10.1038/d41586-019-01985-y.

34. David Feldman, Eric O'Shaughnessy, and Robert Margolis, "Q3/Q4 2019 Solar Industry Update," February 18, 2020, 56. "Solar Market Insight Report 2020 Q2" (Solar Energy Industries Association, June 11, 2020), www. seia.org/research-resources/solar-market-insight-report-2020-q2. Christopher T. M. Clack et al., "Why Local Solar For All Costs Less: A New Roadmap for the Lowest Cost Grid" (Boulder, Colo.: Vibrant Clean Energy, December 1, 2020).

35. Saul Griffith, Sam Calisch, and Laura Fraser, *Rewiring America,* 2020, www.rewiringamerica.org/handbook. Feldman, O'Shaughnessy, and Margolis, "Q3/Q4 2019 Solar Industry Update." "Solar Market Insight Report 2020 Q2." Eric Wesoff, "California's Solar Mandate to Deliver More than 1 GW," *PV Magazine USA,* February 13, 2020, https://pv-mag azine-usa.com/2020/02/13/californias-solar-mandate-to-deliver-more-than-1-gw.

36. Energy Information Administration, "Annual Energy Outlook 2010," 2010. Sarah Zhang, "A Huge Solar Plant Caught on Fire, and That's the Least of Its Problems," *Wired,* May 23, 2016, www.wired.com/2016/05/ huge-solar-plant-caught-fire-thats-least-problems. Chris Martin and Nic Querolo, "A $1 Billion Solar Plant Was Obsolete Before It Ever Went Online," *Bloomberg Businessweek,* January 6, 2020, www.bloomberg.com/news/ articles/2020-01-06/a-1-billion-solar-plant-was-obsolete-before-it-ever-went-online. Amro M. Elshurafa et al., "Estimating the Learning Curve of

Solar PV Balance-of-System for over 20 Countries: Implications and Policy Recommendations," *Journal of Cleaner Production* 196 (September 20, 2018): 122–34, https://doi.org/10.1016/j.jclepro.2018.06.016.

37. "Concentrating Solar Power Projects by Status," National Renewable Energy Laboratory, accessed July 30, 2020, https://solarpaces.nrel.gov/by-status.

38. Energy Information Administration, "Monthly Energy Review." Energy Information Administration, "Geothermal Explained: Use of Geothermal Energy," accessed July 26, 2020, www.eia.gov/energyexplained/geothermal/use-of-geothermal-energy.php.

39. Tim Latimer, Phone interview, December 16, 2019.

40. U.S. Department of Energy, "GeoVision: Harnessing the Heat Beneath Our Feet" (DOE, Office of Science and Technical Information, 2019), www.energy.gov/sites/prod/files/2019/06/f63/GeoVision-full-report-opt.pdf. U.S. Department of Energy, "Hydropower Vision." Dev Millstein, Patrick Dobson, and Seongeun Jeong, "The Potential to Improve the Value of US Geothermal Electricity Generation Through Flexible Operations," *Journal of Energy Resources Technology-Transactions of the ASME* 143, no. 1 (January 1, 2021): 010905, https://doi.org/10.1115/1.4048981.

41. U.S. Department of Energy, "GeoVision: Harnessing the Heat Beneath Our Feet." Tim Latimer, Email, December 17, 2019. Michael J. Coren, "Geothermal Energy, the Forgotten Renewable, Has Finally Arrived," *Quartz*, December 20, 2020, https://qz.com/1947017/geothermal-is-the-electricity-combating-climate-change. Erik Olson, "It's Time to Take Geothermal Energy Seriously," The Breakthrough Institute, September 10, 2020, https://thebreakthrough.org/issues/energy/take-geothermal-seriously. Eli Dourado, "The Biggest No-Brainer in All of Energy Policy," *The Benchmark*, November 2, 2020, https://medium.com/cgo-benchmark/the-biggest-no-brainer-in-all-of-energy-policy-ff4768e6b079.

42. U.S. Department of Energy, "Tapping into Wave and Tidal Ocean Power: 15% Water Power by 2030" (DOE, January 27, 2012), www.energy.gov/articles/tapping-wave-and-tidal-ocean-power-15-water-power-2030. "Mapping and Assessment of the United States Ocean Wave Energy Resource" (EPRI, December 2011), www1.eere.energy.gov/water/pdfs/mappingandassessment.pdf. "Assessment of Energy Production Potential from Tidal Streams in the United States" (Georgia Tech Research Corporation, June 29, 2011). D. S. Jenne and Y.-H. Yu, "Levelized Cost of Energy Analysis of Marine and Hydrokinetic Reference Models," April 2015. "Assessment of Energy Production Potential from Ocean Currents along the United States Coastline" (Georgia Tech Research Corporation, September 15, 2013).

S. Astariz, A. Vazquez, and G. Iglesias, "Evaluation and Comparison of the Levelized Cost of Tidal, Wave, and Offshore Wind Energy," *Journal of Renewable and Sustainable Energy* 7, no. 5 (September 1, 2015): 053112, https://doi.org/10.1063/1.4932154. Eva Segura, Rafael Morales, and José A. Somolinos, "Cost Assessment Methodology and Economic Viability of Tidal Energy Projects," *Energies* 10, no. 11 (November 2017): 1806, https://doi.org/10.3390/en10111806. Laura Castro-Santos et al., "The Levelized Cost of Energy (LCOE) of Wave Energy Using GIS Based Analysis: The Case Study of Portugal," *International Journal of Electrical Power & Energy Systems* 65 (February 1, 2015): 21–25, https://doi.org/10.1016/j.ijepes.2014.09.022. Lazard, "Levelized Cost of Energy and Levelized Cost of Storage—2020."

43. Energy Information Administration, "Annual Energy Outlook 2005," 2005. Gregory Nemet, Email, February 26, 2020.

44. Dr. James H. Williams et al., "Pathways to Deep Decarbonization in the United States," 2014. Jim Williams, Phone interview, November 7, 2019.

45. Williams, interview. The Zero Carbon Consortium, "America's Zero Carbon Action Plan." Larson et al., "Net-Zero America."

46. Christopher Clack, "The Future of Energy: Transmission," https://youtube/E4CJI45rMJM.

47. Energy Information Administration, "Electric Power Monthly," accessed January 14, 2021, www.eia.gov/electricity/monthly/index.php. Data are for twelve months ending October 2020 and include small-scale solar.

48. Guy R. Newsham and Brent G. Bowker, "The Effect of Utility Time-Varying Pricing and Load Control Strategies on Residential Summer Peak Electricity Use: A Review," *Energy Policy*, Large-scale wind power in electricity markets with Regular Papers, 38, no. 7 (July 1, 2010): 3289–96, https://doi.org/10.1016/j.enpol.2010.01.027.

49. "ASCE's 2021 American Infrastructure Report Card | GPA: C-" (American Society of Civil Engineers, 2021), https://infrastructurereportcard.org.

50. Clack, "The Future of Energy: Transmission." Christopher Clack, Phone interview, March 10, 2020.

51. "Report on Barriers and Opportunities for High Voltage Transmissions: A Report to the Committees on Appropriations of Both Houses of Congress" (Federal Energy Regulatory Commission, June 2020), https://cleanenergygrid.org/wp-content/uploads/2020/08/Report-to-Congress-on-High-Voltage-Transmission_17June2020-002.pdf.

52. Vanessa Tutos, In-person interview, October 11, 2018.

53. Xiaodong Du and Ofir D. Rubin, "Transition and Integration of the ERCOT Market with the Competitive Renewable Energy Zones Project," *Energy Journal* 39, no. 4 (July 2018): 235–59.

54. Jennie Jorgenson, Paul Denholm, and Trieu Mai, "Analyzing Storage for Wind Integration in a Transmission-Constrained Power System," *Applied Energy* 228 (October 15, 2018): 122–29, https://doi.org/10.1016/j.apenergy.2018.06.046. Michael Skelly, In-person interview, June 23, 2017. Russell Gold, *Superpower: One Man's Quest to Transform American Energy* (New York: Simon and Schuster, 2019).

55. Peter Behr, "Details Emerge About DOE 'super-grid' Renewable Study," *E&E News*, October 29, 2019, www.eenews.net/stories/1061403455.

56. Christopher T. M. Clack et al., "Evaluation of a Proposal for Reliable Low-Cost Grid Power with 100% Wind, Water, and Solar," *Proceedings of the National Academy of Sciences* 114, no. 26 (June 27, 2017): 6722–27, https://doi.org/10.1073/pnas.1610381114. Clack, interview. Jennifer Weeks, "U.S. Electrical Grid Undergoes Massive Transition to Connect to Renewables," *Scientific American*, accessed July 31, 2020, www.scientificamerican.com/article/what-is-the-smart-grid.

57. Larson et al., "Net-Zero America." The Zero Carbon Consortium, "America's Zero Carbon Action Plan." Patrick R. Brown and Audun Botterud, "The Value of Inter-Regional Coordination and Transmission in Decarbonizing the US Electricity System," *Joule*, December 11, 2020, https://doi.org/10.1016/j.joule.2020.11.013. Robert Walton, "New Transmission Approaches Can Cut Billions in Decarbonization Costs: MIT, Clean Energy Coalition," *Utility Dive*, January 13, 2021, www.utilitydive.com/news/new-transmission-approaches-can-cut-billions-in-decarbonization-costs-mit/593240.

58. "Solving the Climate Crisis: The Congressional Action Plan for a Clean Energy Economy and a Healthy, Resilient, and Just America" (House Select Committee on the Climate Crisis, June 2020), https://perma.cc/P6T4-QKME.

59. Avi Zevin et al., "Building a New Grid Without New Legislation: A Path to Revitalizing Federal Transmission Authorities" (Columbia SIPA Center on Global Energy Policy, December 14, 2020), www.energypolicy.columbia.edu/research/report/building-new-grid-without-new-legislation-path-revitalizing-federal-transmission-authorities. Clack, interview. Eric Wesoff, "Trump Admin Disappears NREL 'Seams' Study and a 250-Ton Chinese Transformer," *PV Magazine USA*, September 23, 2020, https://pv-magazine-usa.com/2020/09/23/morning-brief-trump-admin-disappears-nrel-seams-study-and-a-250-ton-chinese-transformer. FERC, "Report on Barriers and Opportunities for High Voltage Transmissions: A Report to the Committees on Appropriations of Both Houses of Congress."

60. Matthew R. Shaner et al., "Geophysical Constraints on the Reliability of Solar and Wind Power in the United States," *Energy & Environmental Science* 11, no. 4 (April 18, 2018): 914–25, https://doi.org/10.1039/C7EE03029K. Adam Benzion, "Energy Storage—A Trillion-Dollar Holy Grail," *POWER*, February 26, 2020, www.powermag.com/energy-stor age-a-trillion-dollar-holy-grail. David Iaconangelo, "Q&A: Tesla's Ex-Storage Chief on Trump, Musk, and the 'Holy Grail,' " *E&E News*, May 15, 2020, www.eenews.net/stories/1063138901.

61. Wesley J. Cole and Allister Frazier, "Cost Projections for Utility-Scale Battery Storage," June 19, 2019, https://doi.org/10.2172/1529218. Energy Information Administration, "Large Battery Systems Are Often Paired with Renewable Energy Power Plants" (May 18, 2020), www.eia.gov/todayinenergy/detail.php?id=43775. "Battery Pack Prices Fall as Market Ramps Up with Market Average at $156/KWh in 2019," *BloombergNEF* (blog), December 3, 2019, https://about.bnef.com/blog/battery-pack-prices-fall-as-market-ramps-up-with-market-average-at-156-kwh-in-2019. James Temple, "The $2.5 Trillion Reason We Can't Rely on Batteries to Clean up the Grid," *MIT Technology Review*, July 27, 2018, www.technologyreview.com/2018/07/27/141282/the-25-trillion-reason-we-cant-rely-on-batteries-to-clean-up-the-grid. Jesse Jenkins, Phone interview, November 29, 2018.

62. Clack, interview.

63. U.S. Department of Energy, "Hydropower Vision." Energy Information Administration, "Most Pumped Storage Electricity Generators in the U.S. Were Built in the 1970s" (October 31, 2019), www.eia.gov/todayinenergy/detail.php?id=41833.

64. D. S. Tarnay, "Hydrogen Production at Hydropower Plants," *International Journal of Hydrogen Energy* 10, no. 9 (1985): 577–84, https://doi.org/10.1016/0360-3199(85)90032-1.

65. Catherine Morehouse, "Natural Gas Plant Replacing Los Angeles Coal Power to Be 100% Hydrogen by 2045: LADWP," *Utility Dive*, December 12, 2019, www.utilitydive.com/news/natural-gas-plant-replacing-los-angeles-coal-power-to-be-100-hydrogen-by-2/568918. Matthew A. Pellow et al., "Hydrogen or Batteries for Grid Storage? A Net Energy Analysis," *Energy & Environmental Science* 8, no. 7 (2015): 1938–52, https://doi.org/10.1039/C4EE04041D. Tomich, "Gas Plant Developer Bets Big on CO2-Free Hydrogen."

66. Pellow et al., "Hydrogen or Batteries for Grid Storage?" Giacomo Butera, Søren Højgaard Jensen, and Lasse Røngaard Clausen, "A Novel System for Large-Scale Storage of Electricity as Synthetic Natural Gas Using Reversible

Pressurized Solid Oxide Cells," *Energy* 166 (January 1, 2019): 738–54, https://doi.org/10.1016/j.energy.2018.10.079. David Roberts, "The Missing Puzzle Piece for Getting to 100% Clean Power," *Vox*, March 28, 2020, www.vox.com/energy-and-environment/2020/3/28/21195056/renewable-energy-100-percent-clean-electricity-power-to-gas-methane. D. Steward et al., "Lifecycle Cost Analysis of Hydrogen Versus Other Technologies for Electrical Energy Storage" (National Renewable Energy Laboratory, 2009). Tom Brown, Twitter post, August 1, 2020, https://twitter.com/nworbmot.

67. Mark Jacobson, In-person interview, July 31, 2018. Jacobson et al., "Low-Cost Solution to the Grid Reliability Problem with 100% Penetration of Intermittent Wind, Water, and Solar for All Purposes." Mark Z. Jacobson et al., "The United States Can Keep the Grid Stable at Low Cost with 100% Clean, Renewable Energy in All Sectors Despite Inaccurate Claims," *Proceedings of the National Academy of Sciences* 114, no. 26 (June 27, 2017): E5021–23, https://doi.org/10.1073/pnas.1708069114. Clack et al., "Evaluation of a Proposal for Reliable Low-Cost Grid Power with 100% Wind, Water, and Solar."

68. "Race to 100% Clean" (NRDC, June 9, 2020), www.nrdc.org/resources/race-100-clean. Ocasio-Cortez, "Recognizing the Duty of the Federal Government to Create a Green New Deal."

69. Jesse D. Jenkins and Samuel Thernstrom, "Deep Decarbonization of the Electric Power Sector: Insights from Recent Literature" (Energy Innovation Reform Project, March 2017). Jesse D. Jenkins, Max Luke, and Samuel Thernstrom, "Getting to Zero Carbon Emissions in the Electric Power Sector," *Joule* 2, no. 12 (December 19, 2018): 2498–2510, https://doi.org/10.1016/j.joule.2018.11.013. Nestor A. Sepulveda et al., "The Role of Firm Low-Carbon Electricity Resources in Deep Decarbonization of Power Generation," *Joule* 2, no. 11 (November 2018): 2403–20, https://doi.org/10.1016/j.joule.2018.08.006. Alexander E. Macdonald et al., "Future Cost-Competitive Electricity Systems and Their Impact on US CO_2 Emissions," *Nature Climate Change; London* 6, no. 5 (May 2016): 526–31, http://dx.doi.org.ezproxy.rice.edu/10.1038/nclimate2921. Bethany A. Frew et al., "Flexibility Mechanisms and Pathways to a Highly Renewable US Electricity Future," *Energy* 101 (April 15, 2016): 65–78, https://doi.org/10.1016/j.energy.2016.01.079. Mark Z. Jacobson et al., "The United States Can Keep the Grid Stable at Low Cost with 100% Clean, Renewable Energy in All Sectors Despite Inaccurate Claims." Mark Z. Jacobson et al., "Matching Demand with Supply at Low Cost in 139 Countries Among 20 World Regions with 100% Intermittent Wind, Water, and Sunlight (WWS) for All Purposes," *Renewable Energy* 123 (August 2018):

236–48, https://doi.org/10.1016/j.renene.2018.02.009. Jacobson et al., "Low-Cost Solution to the Grid Reliability Problem with 100% Penetration of Intermittent Wind, Water, and Solar for All Purposes." The Zero Carbon Consortium, "America's Zero Carbon Action Plan." Michael Webber, Phone interview, April 7, 2020.

70. "The 2035 Report: Plummeting Wind, Solar, and Battery Costs Can Accelerate Our Clean Electricity Future" (Goldman School of Public Policy, University of California at Berkeley, June 2020), www.2035report.com/downloads.

71. Organisation for Economic Co-operation and Development, "OECD Statistics," accessed July 27, 2020, https://stats.oecd.org.

72. Global Commission on the Geopolitics of Energy Transformation, *A New World: The Geopolitics of the Energy Transformation* (International Renewable Energy Agency, 2019), www.geopoliticsofrenewables.org/assets/geo politics/Reports/wp-content/uploads/2019/01/Global_commission _renewable_energy_2019.pdf.

Chapter 6. Power Shift

1. U.S. Federal Highway Administration, "Moving 12-Month Total Vehicle Miles Traveled" (FRED, Federal Reserve Bank of St. Louis, July 9, 2020), https://fred.stlouisfed.org/series/M12MTVUSM227NFWA. "Transportation Energy Data Book: Edition 38.1" (Oak Ridge National Laboratory, April 30, 2020), https://tedb.ornl.gov/data. "Passenger Travel Facts and Figures 2015" (U.S. Department of Transportation), accessed July 27, 2020, www.bts.dot.gov/sites/bts.dot.gov/files/legacy/PTFF_Complete.pdf.

2. Amber Mahone, Zoom interview, April 10, 2020. Joan Ogden, Zoom interview, May 1, 2020.

3. U.S. Department of Energy, Office of Energy Efficiency and Renewable Energy, "Alternative Fuels Data Center: Hydrogen Fueling Station Locations," accessed July 27, 2020, https://afdc.energy.gov/fuels/hydrogen_locations.html#/find/nearest?fuel=HY&country=US&hy_nonretail=true. Mahone, interview.

4. "Battery Pack Prices Cited Below $100/KWh for the First Time in 2020, While Market Average Sits at $137/KWh," *BloombergNEF* (blog), December 16, 2020, https://about.bnef.com/blog/battery-pack-prices-cited-below-100-kwh-for-the-first-time-in-2020-while-market-average-sits-at-137-kwh. Ranjit R. Desai et al., "Heterogeneity in Economic and Carbon Benefits of Electric Technology Vehicles in the US," *Environmental Science*

& Technology 54, no. 2 (January 21, 2020): 1136–46, https://doi.org/10.1021/acs.est.9b02874. Hanna L. Breetz and Deborah Salon, "Do Electric Vehicles Need Subsidies? Ownership Costs for Conventional, Hybrid, and Electric Vehicles in 14 U.S. Cities," *Energy Policy* 120 (September 1, 2018): 238–49, https://doi.org/10.1016/j.enpol.2018.05.038. "UBS Evidence Lab Electric Car Teardown—Disruption Ahead?" (UBS, May 18, 2017), https://neo.ubs.com/shared/d1wkuDlEbYPjF. Nic Lutsey and Michael Nicholas, "Update on Electric Vehicle Costs in the United States Through 2030" (International Council on Clean Transportation, April 2, 2019), https://theicct.org/sites/default/files/publications/EV_cost_2020_2030_20190401.pdf. Chris Harto, "Electric Vehicle Ownership Costs: Chapter 2—Maintenance" (*Consumer Reports*, September 2020), https://advocacy.consumerreports.org/wp-content/uploads/2020/09/Maintenance-Cost-White-Paper-9.24.20-1.pdf.

5. "Nissan LEAF Generations," *Autolist*, July 10, 2020, https://production-proxy.autolist.com/nissan-leaf/nissan-leaf-generations.

6. Eric A. Taub, "For Electric Car Owners, 'Range Anxiety' Gives Way to 'Charging Time Trauma,'" *New York Times*, October 5, 2017, sec. Automobiles, www.nytimes.com/2017/10/05/automobiles/wheels/electric-cars-charging.html. U.S. Department of Energy, Office of Energy Efficiency and Renewable Energy, "Alternative Fuels Data Center: Electric Vehicle Charging Station Locations," accessed July 27, 2020, https://afdc.energy.gov/fuels/electricity_locations.html#/find/nearest?fuel=ELEC. NACS, "U.S. Convenience Store Count Stands at 152,720 Stores," February 3, 2020, www.convenience.org/Media/Daily/2020/Feb/3/1-US-C-Store-Count-Stands-at-152720-Stores_NACS. Jack Brouwer, Zoom interview, April 30, 2020. Michael Nicholas, Dale Hall, and Nic Lutsey, "Quantifying the Electric Vehicle Charging Infrastructure Gap Across U.S. Markets" (International Council on Clean Transportation, January 2019).

7. Larson et al., "Net-Zero America." The Zero Carbon Consortium, "America's Zero Carbon Action Plan." Trieu T. Mai et al., "Electrification Futures Study: Scenarios of Electric Technology Adoption and Power Consumption for the United States," June 29, 2018, https://doi.org/10.2172/1459351.

8. Charlie Bloch et al., "Breakthrough Batteries: Powering the Era of Clean Electrification" (Rocky Mountain Institute, 2019).

9. James Horrox and Matthew Casale, "Electric Buses in America: Lessons from Cities Pioneering Clean Transportation" (U.S. PIRG Education Fund and Environment America, October 2019). Sven Borén, "Electric Buses' Sustainability Effects, Noise, Energy Use, and Costs," *International*

Journal of Sustainable Transportation, September 17, 2019, 1–16, https://doi.org/10.1080/15568318.2019.1666324. Luis Ignacio Rizzi and Cristobal De La Maza, "The External Costs of Private Versus Public Road Transport in the Metropolitan Area of Santiago, Chile," *Transportation Research Part A: Policy and Practice* 98 (April 1, 2017): 123–40, https://doi.org/10.1016/j.tra.2017.02.002. Marc Prosser, "China's Electric Buses Save More Diesel Than All Electric Cars Combined," *Singularity Hub* (blog), April 22, 2019, https://singularityhub.com/2019/04/22/chinas-electric-buses-save-more-diesel-than-all-electric-cars-combined.

10. Shashank Sripad and Venkatasubramanian Viswanathan, "Performance Metrics Required of Next-Generation Batteries to Make a Practical Electric Semi Truck," *ACS Energy Letters* 2, no. 7 (July 14, 2017): 1669–73, https://doi.org/10.1021/acsenergylett.7b00432.

11. Hengbing Zhao et al., "A Comparison of Zero-Emission Highway Trucking Technologies" (University of California Institute of Transportation Studies, October 18, 2018), https://escholarship.org/uc/item/1584b5z9. Iain Staffell et al., "The Role of Hydrogen and Fuel Cells in the Global Energy System," *Energy & Environmental Science* 12, no. 2 (February 13, 2019): 463–91, https://doi.org/10.1039/C8EE01157E.

12. Zhao et al., "A Comparison of Zero-Emission Highway Trucking Technologies." Andrew Burke and Marshall Miller, "Zero-Emission Medium- and Heavy-Duty Truck Technology, Markets, and Policy Assessments for California" (University of California Institute of Transportation Studies, October 2018), https://escholarship.org/uc/item/7n68r0q8. Dale Hall and Nic Lutsey, "Estimating the Infrastructure Needs and Costs for the Launch of Zero-Emission Trucks" (International Council on Clean Transportation, August 2019). Marshall Miller, Zoom interview, May 21, 2020. Staffell et al., "The Role of Hydrogen and Fuel Cells in the Global Energy System." Joshua S. Hill, "Why Electric Trucks, Not Hydrogen, Will Corner Semi Market and Replace Diesel," *The Driven*, April 29, 2020, https://thedriven.io/2020/04/29/why-electric-trucks-not-hydrogen-will-corner-semi-market-and-replace-diesel. Auke Hoekstra, Twitter direct message, May 23, 2020.

13. Hall and Lutsey, "Estimating the Infrastructure Needs and Costs for the Launch of Zero-Emission Trucks." M. Melaina and M. Penev, "Hydrogen Station Cost Estimates: Comparing Hydrogen Station Cost Calculator Results with Other Recent Estimates" (National Renewable Energy Laboratory, September 2013), https://doi.org/10.2172/1260510.

14. Mahone, interview. Auke Hoekstra, Twitter direct message, May 23, 2020. Ogden, interview.

15. Davis et al., "Net-Zero Emissions Energy Systems."
16. Paul Balcombe et al., "How to Decarbonise International Shipping: Options for Fuels, Technologies and Policies," *Energy Conversion and Management* 182 (February 15, 2019): 72–88, https://doi.org/10.1016/j.enconman.2018.12.080. Andreas Goldmann et al., "A Study on Electrofuels in Aviation," *Energies* 11, no. 2 (February 2018): 392, https://doi.org/10.3390/en11020392. Davis et al., "Net-Zero Emissions Energy Systems."
17. Energy Information Administration, "Residential Energy Consumption Survey 2015" (n.d.), www.eia.gov/consumption/residential/data/2015. Energy Information Administration, "Annual Energy Outlook 2020," 2020. Lucas W. Davis, "What Matters for Electrification? Evidence from 70 Years of U.S. Home Heating Choices" (Energy Institute at Haas, January 2021).
18. Energy Information Administration, "One in Eight U.S. Homes Uses a Programmed Thermostat with a Central Air Conditioning Unit" (July 19, 2017), www.eia.gov/todayinenergy/detail.php?id=32112. Environmental Protection Agency, "Emission Factors for Greenhouse Gas Inventories" (March 9, 2018), www.epa.gov/sites/production/files/2018-03/documents/emission-factors_mar_2018_0.pdf. "Pacific Northwest Pathways to 2050: Achieving an 80% Reduction in Economy-Wide Greenhouse Gases by 2050" (Energy + Environmental Economics, November 2018), www.ethree.com/wp-content/uploads/2018/11/E3_Pacific_Northwest_Pathways_to_2050.pdf. Bill Liss, Phone interview, November 8, 2019. "Bridging the Gap: Gas-Fired Absorption Heat Pumps in America," *CIBSE Journal* (blog), accessed July 28, 2020, www.cibsejournal.com/technical/bridging-the-gap-gas-fired-absorption-heat-pumps. "Greenhouse Gas Emission Reduction Pathways; Phase 1: Gas Technology Pathway Identification" (American Gas Association and Enovation Partners, May 2018), www.aga.org/globalassets/research—insights/reports/ghg-reduction-pathways_phase-1-report.pdf. Edward A. Vineyard, Ahmad Abu-Heiba, and Isaac Mahderekal, "Design and Development of a Residential Gas-Fired Heat Pump," *12th IEA Heat Pump Conference*, 2017, 9.
19. Parth Vaishnav and Adilla Mulia Fatimah, "The Environmental Consequences of Electrifying Space Heating," *Environmental Science & Technology*, July 10, 2020, https://doi.org/10.1021/acs.est.0c02705.
20. Eric D. Lebel et al., "Quantifying Methane Emissions from Natural Gas Water Heaters," *Environmental Science & Technology* 54, no. 9 (05 2020): 5737–45, https://doi.org/10.1021/acs.est.9b07189. Tianchao Hu, Brett C. Singer, and Jennifer M. Logue, "Compilation of Published PM2.5 Emission Rates for Cooking, Candles, and Incense for Use in Modeling of

Exposures in Residences" (Lawrence Berkeley National Laboratory [LBNL], Berkeley, Calif., August 1, 2012), https://doi.org/10.2172/1172959. N. A. Mullen et al., "Results of the California Healthy Homes Indoor Air Quality Study of 2011–2013: Impact of Natural Gas Appliances on Air Pollutant Concentrations," *Indoor Air* 26, no. 2 (2016): 231–45, https://doi.org/10.1111/ina.12190. Kathleen Belanger et al., "Household Levels of Nitrogen Dioxide and Pediatric Asthma Severity," *Epidemiology (Cambridge, Mass.)* 24, no. 2 (March 2013): 320–30, https://doi.org/10.1097/EDE.0b013e318280e2ac. Weiwei Lin, Bert Brunekreef, and Ulrike Gehring, "Meta-Analysis of the Effects of Indoor Nitrogen Dioxide and Gas Cooking on Asthma and Wheeze in Children," *International Journal of Epidemiology* 42, no. 6 (December 1, 2013): 1724–37, https://doi.org/10.1093/ije/dyt150. Wendee Nicole, "Cooking Up Indoor Air Pollution: Emissions from Natural Gas Stoves," *Environmental Health Perspectives* 122, no. 1 (January 2014), https://doi.org/10.1289/ehp.122-A27. Bruce Nilles, Phone interview, May 8, 2020.

21. Sherri Billimoria, Leia Guccione, and Mike Henchen, "The Economics of Electrifying Buildings: How Electric Space and Water Heating Supports Decarbonization of Residential Buildings" (Rocky Mountain Institute, 2018), file:///Users/dsc1/Downloads/RMI_Economics_of_Electrifying_Buildings_2018.pdf. U.S. Department of Energy, "Geothermal Heat Pumps," Energy.gov, accessed July 28, 2020, www.energy.gov/energysaver/heat-and-cool/heat-pump-systems/geothermal-heat-pumps. Ken Eklund and Adria Banks, "Application of Combined Space and Water Heat Pump Systems to Existing Homes for Efficiency and Demand Response: Final Report" (Washington State University Energy Program, September 30, 2017), www.energy.wsu.edu/Documents/Final%20Report%20TIP%20338.pdf.

22. Justin Gerdes, "What Does It Take to Electrify Everything in Your Home?," *GTM*, April 2, 2018, www.greentechmedia.com/articles/read/what-does-it-take-to-electrify-everything-in-your-home. Justin Guay, Email, May 13, 2020.

23. Mahone, interview. Mike Henchen, Phone interview, November 13, 2019.

24. "Equitable Building Electrification: A Framework for Powering Resilient Communities" (The Greenlining Institute, 2019), http://greenlining.org/wp-content/uploads/2019/09/Greenlining_EquitableElectrificationReport_2019_WEB.pdf. Andy Bilich, Michael Colvin, and Timothy O'Connor, "Managing the Transition: Proactive Solutions for Stranded Gas Asset Risk in California" (Environmental Defense Fund, 2019). "California's Gas System in Transition: Equitable, Affordable, Decarbonized,

and Smaller" (Gridworks, 2019), https://gridworks.org/wp-content/up loads/2019/09/CA_Gas_System_in_Transition.pdf. Jeff St. John, "PG&E Gets on Board with All-Electric New Buildings in California," *GreenTech Media*, June 26, 2020, www.greentechmedia.com/articles/read/pge-gets-on-board-with-all-electric-new-buildings-in-california.

25. "Residential Energy Consumption Survey 2015." Michael Colvin, video interview, May 4, 2020.

26. Paul Hope, "Best Induction Cooktops of 2020," *Consumer Reports*, accessed July 28, 2020, www.consumerreports.org/electric-cooktops/the-best-induction-cooktops.

27. Jim Loboy, "Nugget of Knowledge: Cooking with Gas," *WYTV*, February 26, 2018, www.wytv.com/news/daybreak/nugget-of-knowledge-cooking-with-gas. Nilles, interview.

28. Brady Seals and Andee Krasner, "Gas Stoves: Health and Air Quality Impacts and Solutions" (Rocky Mountain Institute, 2020), https://rmi.org/insight/gas-stoves-pollution-health. Lin, Brunekreef, and Gehring, "Meta-Analysis of the Effects of Indoor Nitrogen Dioxide and Gas Cooking on Asthma and Wheeze in Children." Eva Morales et al., "Association of Early-Life Exposure to Household Gas Appliances and Indoor Nitrogen Dioxide with Cognition and Attention Behavior in Preschoolers," *American Journal of Epidemiology* 169, no. 11 (June 1, 2009): 1327–36, https://doi.org/10.1093/aje/kwp067. Jacob Corvidae, On-site visit, August 10, 2018.

29. David Roberts, "Gas Stoves Can Generate Unsafe Levels of Indoor Air Pollution," *Vox*, May 7, 2020, www.vox.com/energy-and-environ ment/2020/5/7/21247602/gas-stove-cooking-indoor-air-pollution-health-risks. "Natural Gas Genius Campaign Planning," American Public Gas Association, November 2019, www.documentcloud.org/documents/6771157-APGA-Genius-Campaign-Phase-4-Planning.html. Zachary D. Weller, Steven P. Hamburg, and Joseph C. von Fischer, "A National Estimate of Methane Leakage from Pipeline Mains in Natural Gas Local Distribution Systems," *Environmental Science & Technology* 54, no. 14 (July 21, 2020): 8958–67, https://doi.org/10.1021/acs.est.0c00437. Colvin, video interview.

30. Mark Z. Jacobson et al., "Impacts of Green New Deal Energy Plans on Grid Stability, Costs, Jobs, Health, and Climate in 143 Countries," *One Earth* 1, no. 4 (December 2019): 449–63, https://doi.org/10.1016/j.oneear.2019.12.003. United Nations Conference on Trade and Development, *Review of Maritime Transport 2019* (Geneva: United Nations, 2019).

31. "ICEF Industrial Heat Decarbonization Roadmap" (Innovation for Cool Earth Forum, December 2019), www.icef-forum.org/pdf/2019/roadmap/ICEF_Roadmap_201912.pdf.

32. S. Julio Friedmann, Zhiyuan Fan, and Ke Tang, "Low-Carbon Heat Solutions for Heavy Industry: Sources, Options, and Costs Today" (Center on Global Energy Policy, October 2019), https://energypolicy.columbia.edu/sites/default/files/file-uploads/LowCarbonHeat-CGEP_Report_100219-2_0.pdf.

33. Mark Ruth, Zoom interview, May 6, 2020. "Secretary Granholm Launches Energy Earthshots Initiative to Accelerate Breakthroughs Toward a Net-Zero Economy," Energy.gov, June 7, 2021, www.energy.gov/articles/secretary-granholm-launches-energy-earthshots-initiative-accelerate-breakthroughs-toward.

34. Lawrence Irlam, "Global Costs of Carbon Capture and Storage: 2017 Update" (Global CCS Institute, June 2017), www.globalccsinstitute.com/archive/hub/publications/201688/global-ccs-cost-updatev4.pdf. Davis et al., "Net-Zero Emissions Energy Systems." S. Julio Friedmann, Zoom interview, April 28, 2020.

35. Friedmann, Fan, and Tang, "Low-Carbon Heat Solutions for Heavy Industry."

36. Trieu Mai, Phone interview, October 31, 2019.

37. Energy Information Administration, "Monthly Energy Review." Mai et al., "Electrification Futures Study."

38. Mark Kane, "350 KW To 1.2 MW Fast Charging with ESS Coming to Norway," *InsideEVs*, May 4, 2019, https://insideevs.com/news/347476/350-kw-12-mw-fast-charging-ess-norway.

39. Thomas A. Deetjen, Andrew S. Reimers, and Michael E. Webber, "Can Storage Reduce Electricity Consumption? A General Equation for the Grid-Wide Efficiency Impact of Using Cooling Thermal Energy Storage for Load Shifting," *Environmental Research Letters* 13, no. 2 (February 2018): 024013, https://doi.org/10.1088/1748-9326/aa9f06.

40. "Impact of Winter Storm Uri on Chemical Markets" (IHS Markit, 2021), https://ihsmarkit.com/topic/impact-of-winter-storm-uri-on-chemical-markets.html.

41. Webber, interview.

42. "The Future of Hydrogen—Analysis" (International Energy Agency, 2020), www.iea.org/reports/the-future-of-hydrogen.

43. "Global Status of CCS 2019" (Global CCS Institute, 2019), www.globalccsinstitute.com/resources/global-status-report. Friedmann, Fan, and Tang, "Low-Carbon Heat Solutions for Heavy Industry." Friedmann, interview. "Carbonomics The Rise of Clean Hydrogen" (Goldman Sachs, 2020), www.goldmansachs.com/insights/pages/gs-research/carbonomics-the-rise-of-clean-hydrogen/report.pdf.

44. J. Robinson, "Cost, Logistics Offer 'Blue Hydrogen' Market Advantages over 'Green' Alternative | S&P Global Platts," *S&P Global Platts*, March 19, 2020, www.spglobal.com/platts/en/market-insights/latest-news/electric-power/031920-cost-logistics-offer-blue-hydrogen-market-advantages-over-green-alternative. Brouwer, interview. "Carbonomics The Rise of Clean Hydrogen." Sivaram, interview.

45. "Hydrogen Pipelines," Energy.gov, accessed July 28, 2020, www.energy.gov/eere/fuelcells/hydrogen-pipelines. Energy Information Administration, "Natural Gas Pipelines," December 5, 2019, www.eia.gov/energy explained/natural-gas/natural-gas-pipelines.php. Brouwer, interview. M. W. Melaina, O. Antonia, and M. Penev, "Blending Hydrogen into Natural Gas Pipeline Networks: A Review of Key Issues" (National Renewable Energy Laboratory, March 2013), www.nrel.gov/docs/fy13osti/51995.pdf.

46. Goldmann et al., "A Study on Electrofuels in Aviation." Robert F. Service, "Ammonia—a Renewable Fuel Made from Sun, Air, and Water—Could Power the Globe Without Carbon," Science | AAAS, July 12, 2018, www.sciencemag.org/news/2018/07/ammonia-renewable-fuel-made-sun-air-and-water-could-power-globe-without-carbon. Balcombe et al., "How to Decarbonise International Shipping."

47. Webber, interview.

48. Energy Information Administration, "Monthly Energy Review."

49. Timothy Searchinger et al., "Use of U.S. Croplands for Biofuels Increases Greenhouse Gases Through Emissions from Land-Use Change," *Science* 319, no. 5867 (February 29, 2008): 1238–40, https://doi.org/10.1126/science.1151861. Jason Hill et al., "Environmental, Economic, and Energetic Costs and Benefits of Biodiesel and Ethanol Biofuels," *Proceedings of the National Academy of Sciences* 103, no. 30 (July 25, 2006): 11206, https://doi.org/10.1073/pnas.0604600103.

50. M. H. Langholtz, B. J. Stokes, and L. M. Eaton, "2016 Billion-Ton Report: Advancing Domestic Resources for a Thriving Bioeconomy," July 6, 2016, https://doi.org/10.2172/1271651. "Corn Production and Portion Used for Fuel Ethanol" (Alternative Fuels Data Center), accessed January 19, 2021, https://afdc.energy.gov/data. Steve Hanson, "Soybean Oil Comprises a Larger Share of Domestic Biodiesel Production" (U.S. Energy Information Administration, May 7, 2019), www.eia.gov/todayinenergy/detail.php?id=39372.

51. "Cellulosic Biofuel Contributions to a Sustainable Energy Future: Choices and Outcomes | Science," accessed July 28, 2020, https://science.sciencemag.org/content/356/6345/eaal2324. "Audi E-Diesel and E-Ethanol—Audi Technology Portal," accessed July 28, 2020, www.audi-technology-

portal.de/en/mobility-for-the-future/audi-future-lab-mobility_en/
audi-future-energies_en/audi-e-diesel-and-e-ethanol.

52. "Biogas Potential in the United States" (National Renewable Energy Laboratory, n.d.), www.nrel.gov/docs/fy14osti/60178.pdf. "Alternative Fuels Data Center: Renewable Natural Gas Production," accessed July 28, 2020, https://afdc.energy.gov/fuels/natural_gas_renewable.html. NREL estimates biogas potential from waste as 420 billion cubic feet per year (www.nrel.gov/docs/fy14osti/60178.pdf), compared to 33,657 bcf dry gas production in 2019 (EIA, "Monthly Energy Review").

53. "Bioenergy and Carbon Capture and Storage" (Global CCS Institute, March 14, 2019).

54. Intergovernmental Panel on Climate Change, *Carbon Dioxide Capture and Storage* (Cambridge: Cambridge University Press, 2005). Elizabeth Connelly, Phone interview, April 24, 2020.

55. "Global EV Outlook 2020—Analysis" (International Energy Agency, June 2020), www.iea.org/reports/global-ev-outlook-2020. Robert Rapier, "Why China Is Dominating Lithium-Ion Battery Production," accessed July 28, 2020, www.forbes.com/sites/rrapier/2019/08/04/why-china-is-dominating-lithium-ion-battery-production/#335a719b3786. Brian Eckhouse, "The U.S. Has a Fleet of 300 Electric Buses. China Has 421,000," *Bloomberg. Com*, May 15, 2019, www.bloomberg.com/news/articles/2019-05-15/in-shift-to-electric-bus-it-s-china-ahead-of-u-s-421-000-to-300. Seamus Garvey, "We Can Decarbonise the UK's Gas Heating Network by Recycling Rainwater—Here's How," The Conversation, accessed July 28, 2020, http://theconversation.com/we-can-decarbonise-the-uks-gas-heating-network-by-recycling-rainwater-heres-how-129497.

56. Sivaram, interview.

Chapter 7. Going Negative

1. Kevin Anderson and Glen Peters, "The Trouble with Negative Emissions," *Science* 354, no. 6309 (October 14, 2016): 182–83, https://doi.org/10.1126/science.aah4567.

2. IPCC, "Global Warming of 1.5°C. An IPCC Special Report on the Impacts of Global Warming of 1.5°C Above Pre-Industrial Levels and Related Global Greenhouse Gas Emission Pathways." Millar et al., "Emission Budgets and Pathways Consistent with Limiting Warming to 1.5 °C." Millar and Friedlingstein, "The Utility of the Historical Record for Assessing the Transient Climate Response to Cumulative Emissions." Matthews et al., "Estimating Carbon Budgets for Ambitious Climate Targets." Corinne

Le Quéré et al., "Global Carbon Budget 2018," *Earth System Science Data* 10, no. 4 (December 5, 2018): 2141–94, https://doi.org/10.5194/essd-10-2141-2018.

3. Hannah Ritchie, "Who Has Contributed Most to Global CO2 Emissions?," *Our World in Data*, October 1, 2019, https://ourworldindata.org/contributed-most-global-co2. Environmental Protection Agency, "Inventory of U.S. Greenhouse Gas Emissions and Sinks: 1990–2019." Most of the scenarios in Larson et al., "Net-Zero America," reach around 1 gigaton per year of new U.S. sinks by 2050, which is not inconsistent with scaling up to 2 gigatons per year later this century.

4. Quéré et al., "Global Carbon Budget 2018." Jens Hartmann et al., "Enhanced Chemical Weathering as a Geoengineering Strategy to Reduce Atmospheric Carbon Dioxide, Supply Nutrients, and Mitigate Ocean Acidification," *Reviews of Geophysics* 51, no. 2 (2013): 113–49, https://doi.org/10.1002/rog.20004.

5. Environmental Protection Agency, "Inventory of U.S. Greenhouse Gas Emissions and Sinks: 1990–2019."

6. National Academies of Sciences, Engineering, and Medicine, *Negative Emissions Technologies and Reliable Sequestration: A Research Agenda* (Washington, D.C.: National Academies Press, 2019), https://doi.org/10.17226/25259.

7. National Academies of Sciences, Engineering, and Medicine, *Negative Emissions Technologies and Reliable Sequestration*.

8. National Academies 2019 grouped biochar with BECCS, which it estimated to have a potential of 0.5 gigaton CO2/year in the United States. A 0.3–2 gigaton CO2/year range globally has been estimated by Sabine Fuss et al., "Negative Emissions—Part 2: Costs, Potentials and Side Effects," *Environmental Research Letters* 13, no. 6 (May 2018): 063002, https://doi.org/10.1088/1748-9326/aabf9f. Jinyang Wang, Zhengqin Xiong, and Yakov Kuzyakov, "Biochar Stability in Soil: Meta-Analysis of Decomposition and Priming Effects," *GCB Bioenergy* 8, no. 3 (2016): 512–23, https://doi.org/10.1111/gcbb.12266.

9. IPCC, "Global Warming of 1.5°C." Leo Hickman, "Timeline: How BECCS Became Climate Change's 'Saviour' Technology," *Carbon Brief*, April 13, 2016, www.carbonbrief.org/beccs-the-story-of-climate-changes-saviour-technology. James H. Williams et al., "Carbon-Neutral Pathways for the United States."

10. Christopher Consoli, "Bioenergy and Carbon Capture and Storage" (Global CCS Institute, 2019), www.globalccsinstitute.com/wp-content/uploads/2019/03/BECCS-Perspective_FINAL_18-March.pdf. National

Academies of Sciences, Engineering, and Medicine, *Negative Emissions Technologies and Reliable Sequestration*. P. A. Turner et al., "The Global Overlap of Bioenergy and Carbon Sequestration Potential," *Climatic Change* 148, no. 1 (May 1, 2018): 1–10, https://doi.org/10.1007/s10584-018-2189-z.

11. Williams, interview.

12. Phil Renforth and Gideon Henderson, "Assessing Ocean Alkalinity for Carbon Sequestration," *Reviews of Geophysics* 55, no. 3 (2017): 636–74, https://doi.org/10.1002/2016RG000533. Energy Information Administration, "Monthly Energy Review."

13. Fuss et al., "Negative Emissions—Part 2." Aaron Strong et al., "Ocean Fertilization: Time to Move On," *Nature* 461, no. 7262 (September 2009): 347–48, https://doi.org/10.1038/461347a. Environmental Protection Agency, "Ocean Dumping: International Treaties," Other Policies and Guidance, U.S. EPA, July 10, 2015, www.epa.gov/ocean-dumping/ocean-dumping-international-treaties.

14. David J. Beerling et al., "Potential for Large-Scale CO_2 Removal via Enhanced Rock Weathering with Croplands," *Nature* 583, no. 7815 (July 2020): 242–48, https://doi.org/10.1038/s41586-020-2448-9.

15. Peter B. Kelemen et al., "Engineered Carbon Mineralization in Ultramafic Rocks for CO_2 Removal from Air: Review and New Insights," *Chemical Geology* 550 (September 20, 2020): 119628, https://doi.org/10.1016/j.chemgeo.2020.119628. Noah McQueen et al., "Ambient Weathering of Magnesium Oxide for CO_2 Removal from Air," *Nature Communications* 11, no. 1 (July 3, 2020): 3299, https://doi.org/10.1038/s41467-020-16510-3. Peter Kelemen, Phone interview, June 26, 2020.

16. John Mason, "Understanding the Long-Term Carbon-Cycle: Weathering of Rocks—a Vitally Important Carbon-Sink," *Skeptical Science*, July 2, 2013, www.skepticalscience.com/weathering.html. Kelemen, interview. Fuss et al., "Negative Emissions—Part 2." National Academies of Sciences, Engineering, and Medicine, *Negative Emissions Technologies and Reliable Sequestration*.

17. National Academies of Sciences, Engineering, and Medicine, *Negative Emissions Technologies and Reliable Sequestration*.

18. National Academies of Sciences, Engineering, and Medicine, *Negative Emissions Technologies and Reliable Sequestration*. Jon Gertner, "The Tiny Swiss Company That Thinks It Can Help Stop Climate Change," *New York Times*, February 12, 2019, sec. Magazine, www.nytimes.com/2019/02/12/magazine/climeworks-business-climate-change.html. Akshat Rathi, "World's First 'Negative Emissions' Plant Turns Carbon

Dioxide into Stone," *Quartz*, October 12, 2017, https://qz.com/1100221/the-worlds-first-negative-emissions-plant-has-opened-in-iceland-turning-carbon-dioxide-into-stone.

19. S. Ong et al., "Land-Use Requirements for Solar Power Plants in the United States" (National Renewable Energy Laboratory, June 1, 2013), https://doi.org/10.2172/1086349. Simon Evans, "Direct CO_2 Capture Machines Could Use 'a Quarter of Global Energy' in 2100," *Carbon Brief*, July 22, 2019, www.carbonbrief.org/direct-co2-capture-machines-could-use-quarter-global-energy-in-2100. Giulia Realmonte et al., "An Inter-Model Assessment of the Role of Direct Air Capture in Deep Mitigation Pathways," *Nature Communications* 10, no. 1 (July 22, 2019): 3277, https://doi.org/10.1038/s41467-019-10842-5. David W. Keith et al., "A Process for Capturing CO_2 from the Atmosphere," *Joule* 2, no. 8 (August 15, 2018): 1573–94, https://doi.org/10.1016/j.joule.2018.05.006. I have assumed from Keith et al. that 5.25 GJ gas is used per ton of CO_2 captured, and that an average U.S. home uses 65 GJ of natural gas per year. Ryan Hanna et al., "Emergency Deployment of Direct Air Capture as a Response to the Climate Crisis," *Nature Communications* 12, no. 1 (January 14, 2021): 368, https://doi.org/10.1038/s41467-020-20437-0.

20. David M. Hart, "Biden's Budget Includes a Jump in Climate Spending–Here's Why Investing in Innovation Is Crucial," *The Conversation*, May 26, 2021, http://theconversation.com/bidens-budget-includes-a-jump-in-climate-spending-heres-why-investing-in-innovation-is-crucial-159506.

21. Catherine Morehouse, "New Mexico Delays 350 MW, 240 MWh Solar+storage Projects Intended to Replace San Juan Coal Plant," *Utility Dive*, May 1, 2020, www.utilitydive.com/news/new-mexico-delays-350-mw-240-mwh-solarstorage-projects-intended-to-replace/577139.

22. Nils Markusson, Duncan McLaren, and David Tyfield, "Towards a Cultural Political Economy of Mitigation Deterrence by Negative Emissions Technologies (NETs)," *Global Sustainability* 1 (2018), https://doi.org/10.1017/sus.2018.10. Duncan McLaren, "Quantifying the Potential Scale of Mitigation Deterrence from Greenhouse Gas Removal Techniques," *Climatic Change* 162 (2020): 2411–28, https://doi.org/10.1007/s10584-020-02732-3. Anderson and Peters, "The Trouble with Negative Emissions."

23. NASA Earth Observatory, "Global Effects of Mount Pinatubo," Text.Article (NASA Earth Observatory, June 15, 2001), https://earthobservatory.nasa.gov/images/1510/global-effects-of-mount-pinatubo.

24. Wake Smith and Gernot Wagner, "Stratospheric Aerosol Injection Tactics and Costs in the First 15 Years of Deployment," *Environmental*

Research Letters 13, no. 12 (November 2018): 124001, https://doi.org /10.1088/1748-9326/aae98d.

25. Peter Irvine et al., "Halving Warming with Idealized Solar Geoengineering Moderates Key Climate Hazards," *Nature Climate Change* 9, no. 4 (April 2019): 295–99, https://doi.org/10.1038/s41558-019-0398-8.

26. Markusson, McLaren, and Tyfield, "Towards a Cultural Political Economy of Mitigation Deterrence by Negative Emissions Technologies."

Chapter 8. Confronting Policy Gridlock

1. Energy Information Administration, "Annual Energy Outlook 2020." "International Energy Outlook 2020" (U.S. Energy Information Administration, October 14, 2020), www.eia.gov/outlooks/ieo.

2. P. A. Geroski, "Models of Technology Diffusion," *Research Policy* 29, no. 4–5 (April 2000): 603–25, https://doi.org/10.1016/S0048-7333(99)00092-X.

3. Mai et al., "Electrification Futures Study." "Tracking Clean Energy Progress" (International Energy Agency, 2020), www.iea.org/topics/tracking-clean-energy-progress. K. Usha Rao and V. V. N. Kishore, "A Review of Technology Diffusion Models with Special Reference to Renewable Energy Technologies," *Renewable and Sustainable Energy Reviews* 14, no. 3 (April 1, 2010): 1070–78, https://doi.org/10.1016/j.rser.2009.11.007.

4. Varun Sivaram, "Pairing Push and Pull Policies: A Heavy-Duty Model for Innovation," Council on Foreign Relations, September 27, 2016, www.cfr.org/blog/pairing-push-and-pull-policies-heavy-duty-model-innovation. Sivaram et al., "To Bring Emissions-Slashing Technologies to Market, the United States Needs Targeted Demand-Pull Innovation Policies." Michael Peters et al., "The Impact of Technology-Push and Demand-Pull Policies on Technical Change—Does the Locus of Policies Matter?," *Research Policy* 41, no. 8 (October 2012): 1296–1308, https://doi.org/10.1016/j.respol.2012.02.004.

5. Robin Kundis Craig, "The Clean Water Act on the Cutting Edge: Climate Change and Water-Quality Regulation," *Natural Resources & Environment* 24, no. 2 (2009): 14–18. Robinson Meyer, "American Oceans Are Becoming More Corrosive. Why Doesn't the EPA Regulate Them?," *The Atlantic*, September 13, 2016, www.theatlantic.com/science/archive/2016/09/why-the-epa-doesnt-regulate-ocean-acidification/499772. Arnold W. Reitze, "Dealing with Climate Change Under the National Environmental Policy Act," *William & Mary Environmental Law and Policy Review* 43, no. 1 (October 2018), https://doi.org/10.2139/ssrn.3404358.

6. Environmental Protection Agency, "Our Nation's Air: Status and Trends Through 2019" (2020), www.epa.gov/air-trends. EPA, "Inventory of U.S.

Greenhouse Gas Emissions and Sinks: 1990–2019." Bachmann, "Will the Circle Be Unbroken."

7. Massachusetts v. EPA, 549 U.S. 497 (2007).

8. Amanda Reilly, "U.S. Court Rejects Obama-Era Plan to Eliminate Some Potent Planet Warming Chemicals," *Science*, August 8, 2017, www.sci encemag.org/news/2017/08/us-court-rejects-obama-era-plan-eliminate-some-potent-planet-warming-chemicals. Joseph Goffman, Phone interview, March 18, 2019. Umair Irfan, "A Federal Court Just Struck down Trump's Attempt to Make Power Plants Even Dirtier," *Vox*, January 19, 2021, www.vox.com/2021/1/19/22239074/affordable-clean-energy-rule-vacated-trump-court-climate-change-obama-biden.

9. President Bill Clinton, The President's Address to a Joint Session of Congress, February 17, 1993. Dawn Erlandson, "The BTU Tax Experience: What Happened and Why It Happened," *Pace Environmental Law Review* 12, no. 1 (September 1994): 173–84. Pomerance, interview.

10. Andrew C. Revkin, "On the Issues: Climate Change—Election Guide 2008," *New York Times*, accessed February 15, 2021, www.nytimes.com/elections/2008/president/issues/climate.html. Nate Allen, *Nancy Pelosi and Newt Gingrich Commercial on Climate Change*, 2008, www.youtube.com/watch?v=qi6n_-wB154&ab_channel=NateAllen. Waxman, "H.R.2454–111th Congress (2009–2010): American Clean Energy and Security Act of 2009."

11. Kate Sheppard, "Everything You Always Wanted to Know About the Waxman-Markey Energy/Climate Bill—in Bullet Points," *Grist*, June 4, 2009, https://grist.org/article/2009-06-03-waxman-markey-bill-break down. Edward Maibach, Phone interview, November 16, 2018. Eric Pooley, *The Climate War: True Believers, Power Brokers, and the Fight to Save the Earth* (New York: Hachette, 2010). Eric Pooley, Phone interview, February 15, 2019.

12. Anthony Leiserowitz et al., "Climate Change in the American Mind: April 2020" (New Haven: Yale University and George Mason University, 2020). Maibach, interview. Justin McCarthy, "Most Americans Support Reducing Fossil Fuel Use," *Gallup*, March 22, 2019, https://news.gallup.com/poll/248006/americans-support-reducing-fossil-fuel.aspx. Alec Tyson and Brian Kennedy, "Two-Thirds of Americans Think Government Should Do More on Climate" (Pew Research Center, June 23, 2020), www.pewresearch.org/science/2020/06/23/two-thirds-of-americans-think-government-should-do-more-on-climate.

13. Anthony Leiserowitz, Phone interview, February 11, 2019. Matto Mildenberger and Dustin Tingley, "Beliefs About Climate Beliefs: The Impor-

tance of Second-Order Opinions for Climate Politics," *British Journal of Political Science* 49, no. 4 (2019): 1279–1307, https://doi.org/10.1017/S0007123417000321. Alexander Hertel-Fernandez, Matto Mildenberger, and Leah C. Stokes, "Legislative Staff and Representation in Congress," *American Political Science Review* 113, no. 1 (February 2019): 1–18, https://doi.org/10.1017/S0003055418000606. Mildenberger, interview.

14. Emily Witt, "The Optimistic Activists for a Green New Deal: Inside the Youth-Led Singing Sunrise Movement," *New Yorker*, December 23, 2018, www.newyorker.com/news/news-desk/the-optimistic-activists-for-a-green-new-deal-inside-the-youth-led-singing-sunrise-movement. "Sunrise Movement," Sunrise Movement, accessed August 12, 2020, www.sunrisemovement.org. "About Citizens' Climate Lobby," Citizens' Climate Lobby, accessed August 12, 2020, https://citizensclimatelobby.org/about-ccl. Mark Reynolds, phone interview, February 12, 2019. "What We Stand For," *RepublicEn* (blog), accessed August 12, 2020, https://republicen.org/about. Bob Inglis, In-person interview, October 3, 2018.

15. Climate Leadership Council, "Our Plan," *Climate Leadership Council* (blog), accessed July 14, 2020, https://clcouncil.org/our-plan. Climate Leadership Council, "Economists' Statement on Carbon Dividends." Baker et al., "The Conservative Case for Carbon Dividends." Katie Worth, "In Shift, Key Climate Denialist Group Heartland Institute Pivots to Policy," *PBS Frontline*, November 2, 2018, www.pbs.org/wgbh/frontline/article/in-shift-key-climate-denialist-group-heartland-institute-pivots-to-policy.

16. "Race to Zero Campaign | UNFCCC," accessed August 17, 2020, https://unfccc.int/climate-action/race-to-zero-campaign. Chris Mooney, "This Is Where Obama's Hugely Ambitious Climate Policies Were Headed—Before Trump Came Along," *Washington Post*, November 16, 2016. David Roberts, "Microsoft's Astonishing Climate Change Goals, Explained," *Vox*, July 30, 2020, www.vox.com/energy-and-environment/2020/7/30/21336777/microsoft-climate-change-goals-negative-emissions-technologies.

17. Theda Skocpol, Phone interview, January 27, 2019. Jerry Taylor, Phone interview, May 1, 2019. Emily Pontecorvo, "One of the Country's Biggest Climate Denier Groups Just Did an About-Face," *Grist*, November 13, 2019, https://grist.org/article/one-of-the-countrys-biggest-climate-denier-groups-just-did-an-about-face. Jennifer Dlouhy, "U.S. Chamber to Re-Examine Climate Policy That Cost It Members," *Bloomberg*, September 24, 2019, www.bloomberg.com/news/articles/2019-09-24/u-s-chamber-to-re-examine-climate-policy-that-cost-it-members. "Manufacturers to

Congress: Act on Climate," NAM, October 1, 2019, www.nam.org/manufac
turers-demand-congress-acts-on-climate-change-6021. Greg Bertelsen,
Phone interview, August 13, 2020. Theda Skocpol, "Naming the Problem:
What It Will Take to Counter Extremism and Engage Americans in the
Fight Against Global Warming," January 2013, https://grist.org/wp-content/
uploads/2013/03/skocpol-captrade-report-january-2013y.pdf. Martin
Durbin, "An Update to the Chamber's Approach on Climate," U.S. Cham-
ber of Commerce, January 19, 2021, www.uschamber.com/series/above-the-
fold/update-the-chambers-approach-climate. Valerie Volcovici, "U.S. CEO
Group Says It Supports Carbon Pricing to Fight Climate Change," *Reuters*,
September 16, 2020, www.reuters.com/article/usa-business-carbonpricing-
idUSKBN2672W4.

18. Leah Cardamore Stokes, *Short Circuiting Policy: Interest Groups and
the Battle Over Clean Energy and Climate Policy in the American States*
(New York: Oxford University Press, 2020). Leah Stokes, Phone interview, November 26, 2018. Cullenward and Victor, *Making Climate Policy
Work*.

19. "Clean Transportation Leaders Launch Zero Emission Transportation
Association," ZETA, November 16, 2020, www.zeta2030.org/news/clean-
transportation-leaders-launch-zero-emission-transportation-association.
Robinson Meyer, "What Donald Trump Taught the Electric-Car Industry," *The Atlantic*, November 17, 2020, www.theatlantic.com/science/
archive/2020/11/what-donald-trump-taught-the-electric-car-industry
/617124.

20. Zachary Pleat, "Right-Wing Media Launch Unhinged Attacks on Greta
Thunberg" (Media Matters, September 24, 2019), www.mediamatters.
org/dinesh-dsouza/right-wing-media-launch-unhinged-attacks-greta-
thunberg. Eve Peyser, "The Right-Wing Media Is Desperately Addicted
to Alexandria Ocasio-Cortez," *Vice*, December 7, 2018, www.vice.com/
en_us/article/qvqeaq/the-right-wing-media-cant-quit-alexandria-ocasio-
cortez. Taylor, interview. Scott Waldman and Mark K. Matthews, "GOP
Criticizes Its Own on Climate," *E&E News*, May 28, 2019, www.eenews.
net/stories/1060410993. Allison Fisher, "Foxic: Fox News Network's
Dangerous Climate Denial 2019" (Public Citizen, April 13, 2019). Steven
Perlberg, "She Tried to Report on Climate Change. Sinclair Told Her to
Be More 'Balanced,'" *BuzzFeed News*, April 22, 2018, www.buzzfeednews.
com/article/stevenperlberg/sinclair-climate-change. Naomi Oreskes and
Erik M. Conway, *Merchants of Doubt: How a Handful of Scientists Obscured
the Truth on Issues from Tobacco Smoke to Global Warming* (New York:
Bloomsbury Publishing USA, 2010).

21. Jacob S. Hacker and Paul Pierson, "Policy Feedback in an Age of Polarization," *The ANNALS of the American Academy of Political and Social Science* 685, no. 1 (September 1, 2019): 8–28, https://doi.org/10.1177/0002716219871222.

22. Danny Cullenward, In-person interview, July 30, 2018.

23. Goffman, interview. Brenda Mallory, Joe Goffman, and Jennifer Macedonia, "Climate 21 Project Transition Memo: Environmental Protection Agency," 2020, https://climate21.org/documents/C21_EPA.pdf.

24. Michael Burger, "Summary: Combating Climate Change with Section 115 of the Clean Air Act" (Sabin Center for Climate Change Law, Columbia Law School, 2020), https://doi.org/10.4337/9781786434616.

25. Columbia SIPA Center on Global Energy Policy, "What You Need to Know About a Federal Carbon Tax in the United States," accessed August 12, 2020, www.energypolicy.columbia.edu/what-you-need-know-about-federal-carbon-tax-united-states.

26. Climate Leadership Council, "Economists' Statement on Carbon Dividends." Baker et al., "The Conservative Case for Carbon Dividends." Noah Kaufman, Phone interview, June 20, 2018. National Academies of Sciences, Engineering, and Medicine, *Accelerating Decarbonization of the U.S. Energy System*.

27. Baker et al., "The Conservative Case for Carbon Dividends."

28. Andrew L. Goodkind et al., "Fine-Scale Damage Estimates of Particulate Matter Air Pollution Reveal Opportunities for Location-Specific Mitigation of Emissions," *Proceedings of the National Academy of Sciences* 116, no. 18 (April 30, 2019): 8775–80, https://doi.org/10.1073/pnas.1816102116. Aaron J. Cohen et al., "Estimates and 25-Year Trends of the Global Burden of Disease Attributable to Ambient Air Pollution: An Analysis of Data from the Global Burden of Diseases Study 2015," *Lancet (London, England)* 389, no. 10082 (May 13, 2017): 1907–18, https://doi.org/10.1016/S0140-6736(17)30505-6. Strasert, Teh, and Cohan, "Air Quality and Health Benefits from Potential Coal Power Plant Closures in Texas."

29. John Horowitz et al., "Methodology for Analyzing a Carbon Tax" (U.S. Department of the Treasury, January 2017). Columbia SIPA Center on Global Energy Policy, "What You Need to Know About a Federal Carbon Tax in the United States." Bob Inglis, interview. James Baker, Phone interview, June 26, 2018. Leiserowitz, interview. Bertelsen, interview.

30. Natalie Fitzpatrick et al., "American Opinions on Carbon Taxes and Cap-and-Trade: 10 Years of Carbon Pricing in the NSEE," *Issues in Energy and Environmental Policy*, no. 35 (June 2018). David Roberts, "Washington Votes No on a Price on Carbon Emissions—Again," *Vox*, November 6,

2018, www.vox.com/energy-and-environment/2018/9/28/17899804/ washington-1631-results-carbon-fee-green-new-deal. Kate Aronoff, "BP Claims to Support Taxing Carbon, but It's Spending $13 Million Against an Initiative That Would Do Just That," *The Intercept*, November 1, 2018, https://theintercept.com/2018/11/01/bp-washington-state-carbon-tax-initiative.

31. Ben Geman, "1 Big Thing: The Push to Shape a Carbon Tax," *Axios*, June 21, 2018, www.axios.com/newsletters/axios-generate-a792273e-ac7d-4828-bd3c-402ff043842d.html?chunk=0#story0. Andrew Duehren, "Bipartisan Group Backs Gas-Tax Increase as Option to Fund Infrastructure," *Wall Street Journal*, April 23, 2021, sec. Politics, www.wsj.com/articles/biparti san-group-backs-gas-tax-increase-to-fund-infrastructure-11619188415.

32. World Bank, "Carbon Pricing Dashboard," accessed August 12, 2020, https://carbonpricingdashboard.worldbank.org. Noah Kaufman, "Congressional Testimony of Noah Kaufman, Ph.D.," Subcommittee on Environment & Climate Change of the Committee on Energy and Commerce (2019), https://energypolicy.columbia.edu/sites/default/files/file-uploads/ Kaufman-Testimony_CGEP_Commentary_120319-2.pdf.

33. Stephen Lee and Ellen M. Gilmer, "Key to Biden Climate Agenda: The Social Cost of Carbon Explained," *Bloomberg Law*, January 21, 2021, https:// news.bloomberglaw.com/environment-and-energy/key-to-biden-climate-agenda-the-social-cost-of-carbon-explained. Dana Nuccitelli, "The Trump EPA Is Vastly Underestimating the Cost of Carbon Dioxide Pollution to Society, New Research Finds," *Yale Climate Connections*, July 30, 2020, https:// yaleclimateconnections.org/2020/07/trump-epa-vastly-underestimating-the-cost-of-carbon-dioxide-pollution-to-society-new-research-finds. White House, "Executive Order on Protecting Public Health and the Environment and Restoring Science to Tackle the Climate Crisis."

34. Nuccitelli, "The Trump EPA Is Vastly Underestimating the Cost of Carbon Dioxide Pollution to Society." "Q&A: The Social Cost of Carbon," *Carbon Brief*, February 14, 2017, www.carbonbrief.org/qa-social-cost-carbon.

35. Van Jones, *The Green Collar Economy: How One Solution Can Fix Our Two Biggest Problems* (New York: HarperCollins, 2009). Ocasio-Cortez, "Recognizing the Duty of the Federal Government to Create a Green New Deal."

36. Ocasio-Cortez, "Recognizing the Duty of the Federal Government to Create a Green New Deal."

37. "Green New Deal," Sunrise Movement, accessed August 13, 2020, www. sunrisemovement.org/green-new-deal. Jessica McDonald, "How Much

Will the 'Green New Deal' Cost?," *FactCheck.Org*, March 14, 2019, www.factcheck.org/2019/03/how-much-will-the-green-new-deal-cost. Dino Grandoni and Felicia Sonmez, "Senate Defeats Green New Deal, as Democrats Call Vote a 'Sham,' " *Washington Post*, accessed August 13, 2020, www.washingtonpost.com/powerpost/green-new-deal-on-track-to-senate-defeat-as-democrats-call-vote-a-sham/2019/03/26/834f3e5e-4fdd-11e9-a3f7-78b7525a8d5f_story.html. Taylor, interview. Lauren Egan, "Trump Lets Loose at CPAC in Longest Speech of His Presidency," *NBC News*, March 2, 2019, www.nbcnews.com/politics/donald-trump/trump-lets-loose-cpac-longest-speech-his-presidency-n978556.

38. Zoya Teirstein, "Poll: The Green New Deal Is as Popular as Legalizing Weed," *Grist*, July 22, 2019, https://grist.org/article/poll-the-green-new-deal-is-as-popular-as-legalizing-weed. Benjamin Sovacool, interview.

39. "ASCE's 2021 American Infrastructure Report Card." National Academies of Sciences, Engineering, and Medicine, *Accelerating Decarbonization of the U.S. Energy System*. Hal Harvey et al., "The Costs of Delay" (Energy Innovation, 2021), https://energyinnovation.org/wp-content/uploads/2021/01/Cost_of_Delay.pdf.

40. Severin Borenstein, In-person interview, August 3, 2018. International Energy Agency, "Tracking Clean Energy Progress."

41. Adele Morris, Phone interview, November 15, 2018.

42. Biden, "Plan for Climate Change and Environmental Justice." National Academies of Sciences, Engineering, and Medicine, *Accelerating Decarbonization of the U.S. Energy System*. Sivaram et al., "To Bring Emissions-Slashing Technologies to Market, the United States Needs Targeted Demand-Pull Innovation Policies."

43. U.S. Department of Energy, "GeoVision: Harnessing the Heat Beneath Our Feet."

44. Fergus Green, Skype interview, October 24, 2019.

45. Joby Warrick and Juliet Eilperin, "Obama Announces Moratorium on New Federal Coal Leases," *Washington Post*, accessed August 14, 2020, www.washingtonpost.com/news/energy-environment/wp/2016/01/14/obama-administration-set-to-announce-moratorium-on-some-new-federal-coal-leases. Dustin Weaver, "Trump Administration Ends Obama's Coal-Leasing Freeze," *The Hill*, March 29, 2017, https://thehill.com/policy/energy-environment/326375-interior-department-ends-obamas-coal-leasing-freeze. Michael Liebreich, "Liebreich: Climate Lawsuits—An Existential Risk to Fossil Fuel Firms?," *BloombergNEF* (blog), September 26, 2019, https://about.bnef.com/blog/liebreich-climate-lawsuits-existential-risk-fossil-fuel-firms.

46. Stanley Reed and Claire Moses, "Shell Must Reduce Emissions, Dutch Court Rules," *New York Times*, May 26, 2021, www.nytimes. com/2021/05/26/business/royal-dutch-shell-climate-change.html. Baker et al., "The Conservative Case for Carbon Dividends." Lee Wasserman and David Kaiser, "Opinion | Beware of Oil Companies Bearing Gifts," *New York Times*, July 25, 2018, sec. Opinion, www.nytimes.com/2018/07/25/ opinion/carbon-tax-lott-breaux.html.

47. " 'We Are Still In' Declaration," accessed August 14, 2020, www.wearestil lin.com/we-are-still-declaration. Julia Pyper, "Tracking Progress on 100% Clean Energy Targets," *GTM*, November 12, 2019, www.greentechmedia. com/articles/read/tracking-progress-on-100-clean-energy-targets. "Committed," Sierra Club, June 18, 2020, www.sierraclub.org/ready-for-100/commitments.

48. "Partisan Composition of State Legislatures," Ballotpedia, accessed August 14, 2020, https://ballotpedia.org/Partisan_composition_of_state_ legislatures.

49. Tabuchi, "New Rule in California Will Require Zero-Emissions Trucks." Jesse McKinley and Brad Plumer, "New York to Approve One of the World's Most Ambitious Climate Plans," *New York Times*, June 18, 2019, www.nytimes.com/2019/06/18/nyregion/greenhouse-gases-ny.html. David Roberts, "A Closer Look at Washington's Superb New 100% Clean Electricity Bill," *Vox*, April 18, 2019, www.vox.com/energy-and-environ ment/2019/4/18/18363292/washington-clean-energy-bill. Charles F. Sabel and Jonathan Zeitlin, "Experimentalist Governance," in *The Oxford Handbook of Governance* (Oxford: Oxford University Press, 2012), 169–83, https://doi.org/10.1093/oxfordhb/9780199560530.013.0012. National Academies of Sciences, Engineering, and Medicine, *Accelerating Decarbonization of the U.S. Energy System*.

ACKNOWLEDGMENTS

I am deeply indebted to the nearly one hundred people who gave their time to be interviewed for this book. Although these individuals are too numerous to list here, their insights repeatedly challenged my thinking and informed all that is best in this book. Especially influential were interviews with and suggestions from David Victor, Katharine Hayhoe, Melinda Kimble, Jesse Jenkins, Jim Williams, and Amory Lovins. Michael Webber graciously provided the foreword for this book.

Several colleagues at Rice University provided invaluable feedback on portions of this book: Elizabeth Festa, Tracy Volz, and Leonardo Dueñas Osorio. At Yale University Press, my initial editor, Joseph Calamia, shepherded the proposal to approval and guided the first three chapters. Editor Jean Thomson Black and her assistant Elizabeth Sylvia carried it from there. Any mistakes that remain are my own.

This project is the culmination of three decades of exploration, kicked off with an independent research project in the middle school class of Ruth Ann Dixon. Along the way, I was mentored as a writer by C. Dow Tate and as an atmospheric scientist by Daniel Jacob and Ted Russell. Rice University provided me the academic freedom to write this book, including a one-semester sabbatical to begin interviews and research for it.

Above all, I have been blessed by the love, support, and patience of my children, Mira and Jacob; my parents, Bob and Joni; and my wife, Shubhada Hooli.

INDEX

acidification: of oceans, 131–32, 138; of rain, ix, 17; of soil, 132

Africa, 101, 125

agriculture, xiv, 122; climate change vulnerability of, 7, 8; emissions from, 57, 58, 71, 159; livestock emissions in, 49, 57, 71, 72; negative emissions from, 126, 128–30; slash-and-burn, 49

air conditioning, 72, 82, 92, 103, 112, 117

air pollution, 112, 122, 126, 142, 153; acid rain linked to, ix; from gas cooking, 115; as public health hazard, ix, xiii, 15, 71, 108, 115; from smog, 33, 70, 150

Aklin, Michael, 40

algae, 71, 122, 123, 132

Allam cycle, 81, 150

Alliance of Small Island States, 32

Amazon.com, 61, 136

American Clean Power Association, 148

American Council for an Energy Efficient Economy (ACEEE), 64

American Gas Association (AGA), 114

American Legislative Council, 146

American Petroleum Institute, 147

American Society of Civil Engineers, 93, 157

America's Pledge, 48

ammonia, 71, 116, 121, 129

anaerobic digesters, 71

Antarctica, 13, 19, 82

Apollo Program, 61, 99

appliances, 66, 68–69, 103, 117, 140

ARPA-E (Advanced Research Projects Agency—Energy), 96, 101

Asheim, Geir, 53, 55

Asian financial crisis, 24

asthma, ix, 115

Audi, 123

Australia, 55, 86

aviation, 68, 110–11, 123, 126, 131

Axelrod, Robert, 37–41, 43, 45

Baker, James, 146, 153

Ban Ki-moon, 12

Barrett, Amy Coney, 151

batteries, 98; lithium-ion, 60, 96, 105, 107, 108, 124

Bell Telephone Laboratories, 83, 84

Benny, Jack, 114

Berkeley, Calif., 114

Berlin Mandate, 22

Bertelsen, Greg, 147, 153

Biden, Joseph R., 41, 154–55, 156; clean energy goals of, 35, 61, 73, 99, 136, 158; drilling on public lands paused by, 55, 159; Paris Agreement rejoined by, 30; procurement policies of, 62, 155

Biniaz, Susan, 16, 36, 47

biochar, 129–30, 135

biodiesel, 122

biofertilizer, 71

biofuels, 9, 72, 92, 103, 110, 121, 122–24, 159

biomass energy with carbon capture and storage (BECCS), 124, 130–31, 134, 135

Bismarck, Otto von, 163

Bodansky, Daniel, 26

Bolt electric vehicle, 105–6

Borenstein, Severin, 158

Brazil, 2

Breakthrough Energy Ventures, 61, 88, 101

Brexit, 43

British Petroleum (BP), 136, 146

Brouwer, Jack, 106, 120, 121

Brown, Tom, 98

building codes, 1, 41, 68, 86

buses, 107, 108, 126

Bush, George H. W., 15, 17, 18, 19, 21

Bush, George W., xiii, 24–25, 62, 81

Business Roundtable, 147

Byrd, Robert, 22

calcium, 132, 133

calcium carbonate, 132

California, 71, 84, 161–62; building codes in, 41, 68, 86; emission standards in, 67–68; fuel cells trucks in, 109; geothermal energy in, 87, 88; net-zero goal in, 113–14; solar energy in, 68, 86, 87

Callahan, Kateri, 66

Calpine, 146

Canada, 26, 124

cancer, 15, 108

cap and trade, 143–44; for acid rain, 17; in Clean Air Act, 142; in Europe, 42; in Kyoto Protocol, 23, 27; in Waxman-Markey bill, 41, 144, 146–47, 151

carbon capture, 50, 60, 73, 129, 134, 142; in agriculture, 129–31; in biomass energy production, 123–24; from coal and gas, 79–82; cost of, 90; electricity used in, 99; energy efficiency vs., 153; through engineered weathering, 132–33; through forestation, 128; in hydrogen production, 119, 120; for industrial use, 116–17; for ships and planes, 110–11; storing product of, 9, 33, 49, 80, 101, 102, 119, 131; synthetic fuels and, 121

carbon dioxide, 31, 32, 52, 123; atmospheric persistence of, ix, xiii, 1, 8, 138; fossil fuels linked to, 15–16, 20, 57, 105; as global problem, xiii, 1, 13, 20, 26; as hydrogen production byproduct, 120; industry and, 5, 116–17; oceans acidified by, 33; other greenhouse gases vs., 33, 35, 69; outlook for, 9; pervasiveness of, 15–16, 35; in shipping and aviation emissions, 110; temperature sensitivity to, 6, 13, 127; U.S. emissions of, 9, 16, 57, 103; in vehicle emissions, 103

Carbon Engineering, 134–35

carbon monoxide, 112, 142

carbon sinks, 8–9, 33, 58, 64, 126–39, 159

carbon tariffs, 41, 42, 43, 49

cellulose, 123

cement, 57, 103, 115, 116, 117
Chase, Jenny, 85
Cheney, Dick, 25, 26
Chevrolet Bolt, 105–6
China, 1; coal mining in, 55;
electric vehicles in, 108, 124;
emissions in, xiv, 1–2, 3, 21,
22, 26, 27, 29, 30, 42; net-zero
commitment by, 43, 46; renewable energy in, 84, 101
chlorine, 13
chlorofluorocarbons (CFCs), 20,
69, 164; ozone depletion linked
to, 13, 33, 72, 126; phaseout of,
13, 14–15, 17–19, 21, 35, 41, 47,
51–52, 72; potency of, 5, 15;
sources of, 15
Christensen, Dane, 69
Citizens' Climate Lobby, 146
Clack, Christopher, 91, 93, 95
Clean Air Act (1963), 15, 17,
142–43, 150
Clean Development Mechanism,
23
Clean Line Energy, 94, 95
Clean Power Plan, 142, 150, 151,
161
Clean Water Act (1972), 142
Climate Action Network, 29
Climate Action Tracker, 34, 35
climate clubs, 49–56, 117, 125, 141,
164
Climate Leadership Council, 146,
152, 161
Climate Pledge Fund, 61
Climate Mayors Electric Vehicle
Purchasing Collaborative, 62
climate vs. weather, 4
The Climate War (Pooley), 144
Climeworks, 134

Clinton, Bill, 21, 23–24, 143
coal, ix, 22, 43, 111; carbon capture
from, 79–80, 83, 90, 134; in
China, 43, 55, 101; continuous
output of, 91; emissions from,
54, 80–81; leaks from, 57, 70,
71; in natural gas production,
119; oil and gas vs., 10, 35,
70, 73; outlook for, 99, 164;
political opposition to, 53; rail
transport of, 115
Colvin, Michael, 114, 115
Competitive Enterprise Institute,
24, 148
Connelly, Elizabeth, 124
Conference of the Parties (COP),
18, 29, 30
Conway, Erik, 148
cooking, 92, 112, 114–15
cooperation, in game theory,
37–45
Copenhagen Accord, 26–28, 30–31
coral, 4, 7, 131, 132
corn, 122, 129
Corvidae, Jacob, 115
cover crops, 129, 135
COVID-19 pandemic, 35, 44, 80,
88, 103, 117
Crescent Dunes, Nev., 78, 87
Cullenward, David, 147, 149
cyclones, 138
Cyclotron Road, 61, 88

Daimler, 109
dairy farms, 122, 123
DAYS (Duration Addition to
ElectricitY Storage), 96–97
daytime vs. nighttime temperature, 5
DDT, 63

deforestation, 49
Denmark, 82, 101
developing nations, 19–24, 27–28,
    34, 43; clean energy technology
    for, 51, 52–53
diesel fuel, ix, 64, 67, 77, 105,
    107–8, 121, 122
diffusion of technology, 59
diplomacy: American strength in,
    2, 3; domestic policy linked to,
    3; game theory and, 37–40
Direct Air Carbon Capture and
    Storage (DACCS), 134–35
discounting, of energy savings, 65
drought, x, 1, 7, 8, 138
Du Pont, 15

Earth Day (1970), 12
Earth Summit (1992), 17–18
Eavor Technologies, 88
economic impacts, 7, 31–32
electricity: decarbonization goal
    for, 73, 90–102, 117–19; growing
    demand for, 117–18; in industry,
    116–17; sources of, 74, 83; uneven
    use of, 103; vehicles powered
    by, 35, 58, 62, 64, 69, 72, 91, 99,
    103–11, 118, 124, 126, 141, 142, 148
El Niño, 16
emissions standards, 1, 150
emissions taxes, 62, 151–54, 163
energy efficiency, 35, 50, 57–58,
    63–69, 126, 136; carbon capture
    vs., 153; clean electricity linked
    to, 99; of heat pumps, 64, 99,
    111, 112, 113, 119; of homes and
    businesses, 62; supply curves
    for, 64–65; vehicular, 65, 140
Energy Information Administra-
    tion (EIA), 35, 90

Energy Policy Act (2005), 76, 79,
    90, 95–96
Energy Policy and Conservation
    Act (1975), 66
Energy Star program, 67
engineered weathering, 132–33
Environmental Defense Fund,
    70, 145
environmental justice, 126,
    136–37
Environmental Protection Agency
    (EPA), 15, 27, 70, 115, 150–51;
    airplane emission standards set
    by, 68; biofuels and, 123; CFCs
    banned by, 13; greenhouse
    gases regulated by, 142
ethanol, 117, 122, 123, 130
European Green Deal, 42
European Investment Bank, 52
European Union, 1, 3, 23, 26, 42,
    67, 124, 162
Evans, Ben, 63, 69
Evans, Simon, 45
The Evolution of Cooperation (Axel-
    rod), 38
Exelon, 146
ExxonMobil, 123, 146

Fabius, Laurent, 12, 27, 29
Federal Energy Regulatory Com-
    mission (FERC), 95
feedback cycles, 6
fertilizer, 123, 128, 131, 135; ingre-
    dients for, 97, 116, 119, 129;
    nitrous oxide linked to, 122; for
    oceans, 132; soils nitrogenized
    by, 71, 72, 133
Fervo Energy, 88, 92
Field Observatory for Research in
    Geothermal Energy, 88–89

Finland, 78
fires, x, 1, 7, 33, 49, 128
floods, x, 1, 7, 138
Ford, Michael, 78–79
Ford Motor Company, 146
foreign aid, 17, 18, 21, 27, 41
forests, 7, 33, 49, 128, 131
Fox News, 148
fracking (hydraulic fracturing), 53, 76, 101, 134, 160
France, 46, 70, 101
Freedom Works, 68–69
free-riding, 40–41, 43, 48, 59
Friedmann, Julie, 117, 120
fuel oil, 111, 112
fusion energy, 89, 91
FutureGen project, 81

Gaetz, Matt, 148–49, 154
game theory, 37–45, 46, 47
gas furnaces, 111
gas prices, 68, 152
gas turbines, 91, 98
Gates, Bill, 61, 62, 77, 117
Geden, Oliver, 32–33
Geman, Ben, 154
General Motors, 67, 146
geoengineering, 137–39
geothermal energy, 50, 60, 87–89, 91, 92, 93, 101–2, 159
Germany, 84, 101
Gilbert, Alex, 78
Gillingham, Kenneth, 66
Gingrich, Newt, 144
glaciers, 6
globalization, ix
Global Warming Gridlock (Victor), 50
Goffman, Joseph, 150
Gold, Michael, 94

Gorbachev, Mikhail, 16
Gore, Al, 21, 23, 24
Great Depression, 75
Great Recession, 27–28, 35, 84, 155
Green, Fergus, 159
Green, Jessica, 53
Green Climate Fund, 41, 46, 52
GreenFire Energy, 88
Greenland, 7
Green New Deal, 41, 99, 145, 148, 155–57
"green paradox," 54
grids, x, 1, 91–102
Guay, Justin, 112

Hacker, Jacob, 149
Hagel, Chuck, 22, 25
Hansen, James, 16
Harstad, Bård, 55
hay, 129
Hayhoe, Katharine, 8
Heartland Institute, 146, 148
heating, of homes and businesses, 72, 95, 111–15; electrification of, 58, 91, 98, 100, 103, 118–19; insulation and, 65; peak demands for, 97, 113
heat pumps, 85, 117, 118, 125; cost of, 115, 126; energy efficiency of, 64, 99, 111, 112, 113, 119
heat waves, x, 4, 7, 16, 97
Heliogen, 117
Henchen, Mike, 113
Heritage Foundation, 24
Hickman, Leo, 130
Hoekstra, Auke, 109, 110
Hope, Bob, 114
horizontal drilling, 87–88, 101, 159
Hovi, Jon, 50
hurricanes, x, xiii, 1, 6–7, 8

Hussein, Saddam, 17
hybrid vehicles, 1–4
hydraulic fracturing (fracking), 53, 76, 101, 134, 160
hydrofluorocarbons (HFCs), 5, 52, 72, 141, 143, 164
hydrogen, 58, 85, 103; distribution of, 120–21, 141; fuel cells using, 72, 105, 108–9, 111, 116; "grey" vs. "green" vs. "blue," 97–98, 109, 119–20, 121, 141; storage of, 97; vehicles powered by, 67
hydropower, 73–75, 81–82, 89, 91, 93, 97, 101
Hyundai Motor Company, 109

India, 1, 55, 125
Indonesia, 49
Industrial Revolution, 10
inequality, 7, 50, 107
Infrastructure Report Card, 93, 157
Inglis, Bob, 146, 153
innovation vs. invention, 59
insulation, 62, 64, 65, 111
integrated gasification combined cycle (IGCC), 81
Interconnections Seam Study, 94–95
Intergovernmental Panel on Climate Change (IPCC), 17–18, 33, 79, 127, 130
International Civil Aviation Organization, 51, 68
International Energy Agency, 57, 141, 158
International Energy Conservation Code, 68
International Maritime Organization, 51

International Solar Alliance, 50
invention vs. innovation, 59
iterated (repeated) games, 44–45
irrigation, 131
Ivanpah Solar, 87

Jacobson, Mark, 99
Japan, 2, 26, 84, 101, 124
Jenkins, Jesse, 89, 96
JetBlue, 136
jet fuel, ix, 121
Jones, Van, 155

Kaufman, Noah, 152
Kavanaugh, Brett, 143
Keith, David, 85
Kelemen, David, 133
Kennedy, John F., 99
Kerry, John, 22, 52
Keystone XL pipeline, 55
Kigali Amendments, 52, 72
Kimble, Melinda, 23–24, 29, 30, 44, 56
Koch brothers, 144
Kyoto Protocol, 18, 31, 47, 48; developing countries' misgivings about, 23–24, 43; efforts to amend, 26, 28; impact of, 25–26, 35; U.S. opposition to, 21–25, 30, 41, 42

landfills, 71, 122, 123, 136
land vs. ocean temperature, 4–5
Latimer, Tim, 88–89
leaded gasoline, 63, 164
Leaf electric vehicle, 106
LEDs (light-emitting diodes), 60, 66, 68
Leiserowitz, Anthony, 145, 153
Lenton, Tim, 8

Levin, Kelly, 34
Lewinsky, Monica, 24
lime, 131–32
lithium-ion batteries, 60, 96, 105, 107, 108, 124
livestock emissions, 49, 57, 71, 72
London Amendments, 17, 18, 19
London Protocol on Marine Pollution, 132
loss aversion, 138
Lovering, Jessica, 77
Lovins, Amory, 61, 65
Lyman, Edwin, 78

magnesium, 132, 133
magnesium oxide, 133
Mahouse, Amber, 105, 110, 113
Mai, Trieu, 117
Maibach, Edward, 144, 145
Making Climate Policy Work (Cullenward and Victor), 147
mandates and standards, 60, 164; for appliances, 66, 68–69; for fuel efficiency, 123; innovation in response to, 32; in international agreements, 12, 18, 21, 23, 26, 28, 30, 49; net-zero goal linked to, 63; perils of, 99–100
Manhattan Project, 61
manufacturing, 103, 115–17
Markey, Edward, 144, 155
Massachusetts v. EPA (2007), 27, 142, 151
Maxwell, Ian, 84
McCain, John, 144
McConnell, Mitch, 156
McKinsey and Company, 64
Merchants of Doubt (Oreskes and Conway), 148, 160
mercury emissions, 164

Merkel, Angela, 21
methanation, 121
methane, 72, 134; as biogas component, 123; in hydrogen synthesis, 119; hydropower linked to, 74; leaks of, 57, 70, 71, 112, 115, 126, 141, 142, 164; potency of, 5, 69–70; smog linked to, 33; synthetic, 121–22
methanol, 121
Mexico, 47
Microsoft, 136
Mildenberger, Matto, 10–11, 40, 86, 145, 148
Miller, Marshall, 109
miscanthus, 131
Mississippi Power, 81
mitigation deterrence, 138, 139
Molina, Mario, 13
Montreal Protocol, 41, 47; effectiveness of, 13, 15, 21, 30, 52; irreplicability of, 16; Kigali Amendments to, 52, 72; Kyoto Protocol likened to, 31; London Amendments to, 17, 18, 19; top-down approach of, 17, 30, 35
Moore, David, 84, 85
moral hazard, 137
Morocco, 47
Morris, Adele, 158
Mount Pinatubo, 137
Murkowski, Frank, 22
Murray, Patty, 22

Nadel, Steven, 64
NASA (National Aeronautics and Space Administration), 13
National Academies of Sciences, Engineering and Medicine, 9, 61, 128–31, 133, 135, 157, 162

National Association of Manufacturers, 143, 147
National Environmental Policy Act (1969), 89, 142
National Renewable Energy Laboratory (NREL), 84, 85, 94, 106, 117–18
National Research Conference, 13
natural gas, ix, 53, 134; in Allen cycle, 81; cost of, 68, 73, 76; in hydrogen production, 105, 119–20; methane leaks from, 70, 112; in space and water heating, 111; storage of, 98; wood burning and coal superseded by, 10, 35, 73
Natural Resources Defense Council, 145
negative emissions, 58–59, 72, 99, 141; from agriculture, 126, 128–30; from bioenergy, 123–24; sinks and, 8–9, 33, 58, 64, 126–39; tax policy and, 152
Nemet, Gregory, 90
Nest thermostats, 69, 124
NET Power, 81, 150
net-zero goal: by businesses, 146–47; in California, 113–14; for carbon vs. all greenhouse gases, 33; Chinese pledge of, 43; electric vehicles and, 106–7; in Europe, 26; as Green New Deal goal, 156; mandates and standards linked to, 63; in Paris Agreement, 1, 11, 12, 18, 29, 126; timing of, 8–9, 34; transmission capacity linked to, 95
Nevada, 87
New Energy Nexus, 61
Newsmax, 148

nighttime vs. daytime temperature, 5
Nilles, Bruce, 112, 114
Nissan Leaf, 106
nitrates, 129
nitrogen oxides, 71, 112
nitrous oxide, 5, 69, 72, 141; agricultural emissions of, 57, 58, 71, 122, 129; potency of, 70; sinks lacking for, 33, 126
noise, 108
Nordhaus, William, 31, 49, 50
"no-regret" efficiency, 64
North American Free Trade Agreement (NAFTA), 50
Norway, 48–49, 101
nuclear energy, 60, 72, 93; accidents involving, 75–76; cost overruns and, 87, 90; declining use of, 73, 75–77; in France, 101; from fusion, 89, 91; outlook for, 81–82, 91; public views of, 75, 92; small modular reactors (SMRs) and, 77, 78, 79; waste from, 78–79
NuScale, 77–78

Obama, Barack, 26–27, 62, 142, 154; backlash against, 161; at Beijing summit, 29, 46–47; cap and trade backed by, 143–44; carbon capture goal of, 81; energy efficiency and clean energy policies of, 35, 67, 146; Keystone pipeline blocked by, 55, 159; Paris Agreement backed by, 12
Ocasio-Cortez, Alexandria, 148, 155
ocean currents, 1, 89, 91
ocean sinks, 131–32

ocean vs. land temperature, 5
Ogden, Joan, 105, 110
oil, ix, 6, 53, 88; pipelines for, 55, 102, 155, 159–60, 164; wood burning and coal superseded by, 10
Oklo Inc., 77, 78
Olson, Erik, 89
O'Neill, Paul, 24–25
Oreskes, Naomi, 148
Organisation for Economic Co-operation and Development (OECD), 19
Organization of Oil Exporting Countries (OPEC), 54
ozone depletion, 13–15, 19, 30, 33, 51–52, 126
oxyfuel combustion, 81

Paris Agreement (2015), 28–36, 140, 141, 145, 150, 162; conditional commitments under, 47, 52; Copenhagen shortcomings linked to, 28; as framework, 44–48; insufficiency of, 9; net-zero goal of, 1, 11, 12, 18, 29, 126; oil exporters accommodated in, 48; Rio Treaty linked to, 19; rulebook of, 44; stocktakes in, 44, 45–46; temperature targets in, 4; United States and, 1, 2–3, 18, 30, 34–35, 41, 42, 46, 48
particulates, 5, 6, 142, 150
Pelosi, Nancy, 144, 145
permafrost, 7
perovskites, 85, 159
pesticides, 63
Petra Nove project, 80
photosynthesis, 122, 124, 128, 129, 130, 132

photovoltaics, 84–87, 90
phytoplankton, 132
pickup trucks, 67, 103
Pierson, Paul, 149
Piggot, Georgia, 55
pests, 1, 33, 128
Plant Kemper, 81
Pomerance, Rafe, 11, 24, 143
Pooley, Eric, 144
Powell, Colin, 24
power grids, x, 1, 91–102
power plants, 1, 33, 91, 101, 111, 136; biomass-powered, 130–31, 135; emissions from, 24, 25, 57, 73, 80, 142, 150; fossil fuels used in, 57; natural gas–powered, 81; northeastern states' regulation of, 41–42
prisoner's dilemma, 37–41
Prius hybrid, 104
procurement, by public sector, 62, 155
propane, 111
pulp and paper industry, 116, 130
Purvis, Nigel, 24
pyrolysis, 129

Race to Zero, 146
railways, ix, 68
rainfall, xiii, 6–7, 138; acidic, ix, 17
Rapoport, Anatol, 38, 39, 45, 46, 47
Ravikumar, Arvind, 50
Reagan, Ronald, 15, 16
rebound effects, 66
refining, ix, 103
refrigeration, 72, 103
regulation, 149–51
repeated (iterated) games, 44–45
RepublicEn, 146, 153

retaliation, in game theory, 38–39,
    41–43, 46, 47
Reynolds, Mark, 146
Rice, Condoleezza, 25
rice cultivation, 71
Rio Treaty (United Nations
    Framework Convention on
    Climate Change; UNFCCC),
    18–21, 30, 32, 48, 146
Rocky Mountain Institute (RMI),
    64, 65
Roberts, John G., 151
Rowland, F. Sherwood, 13
Russia, 1–2
Ruth, Mark, 116

Sarkozy, Nicolas, 42
Saudi Arabia, 32, 48
sawmills, 131
Scalia, Antonin, 151
sea ice, 6, 7
sea level rise, 1, 7
shale extraction, 35
Shell, 136, 146, 160
shellfish, 131
shipping, 110–11, 115–16, 123, 126,
    131
Short-Circuiting Policy (Stokes), 147
Shultz, George, 146
Sierra Club, 145, 161
sinks, for carbon emissions, 8–9,
    33, 58, 64, 126–39, 159
Sinn, Hans-Werner, 54
Sivaram, Varun, 3, 85, 120, 124
Skelly, Michael, 94
Skocpol, Theda, 147
slash-and-burn agriculture, 49
small modular reactors (SMRs),
    77, 78, 79
smog, 33, 70, 150

solar energy, 35, 50, 79, 118, 135,
    155; in California, 68, 86, 87;
    declining cost of, 60, 73, 82,
    83–84, 90, 91, 141; photovol-
    taic, 84–87, 90; rooftop, 65;
    variability of, 72, 89, 91, 92,
    96–97, 131
solar geoengineering, 137–39
South Korea, 26
soybean oil, 122
SpaceX, 70
Sovacool, Benjamin, 41, 157
sport utility vehicles, 67, 103
Steffen, Will, 8
steelmaking, 63, 103, 115, 116, 117
Stern, Todd, 30
Stockholm conference (1972),
    12–13, 18
Stokes, Leah, 145, 147
storage: of captured carbon, 9,
    33, 49, 80, 101, 102, 119, 131; of
    electricity, 96–98; of hydrogen,
    97; of natural gas, 98
sulfur dioxide, 80, 142
Sunrise Movement, 145–46, 155,
    156
Superpower (Gold), 94
Sweden, 130
Sweet, William, 27, 28, 45
switchgrass, 131
synthetic fuels, 111, 121–22
Syzygy Plasmonics, 119

Taming the Sun (Sivaram), 85
tax credits: for carbon sequestra-
    tion, 136, 142; for efficiency and
    renewables, 35, 88, 142; for new
    cars, 107, 142
Taylor, Jerry, 147, 148–49, 156
Tea Party, 144, 161

technology, phases of, 59
TerraPower, 77, 92
Tesla Inc., 69, 106, 124
thermostats, 65, 69, 92, 111, 119, 124
Third Derivative, 61
Three Mile Island accident (1979), 75
Thunberg, Greta, 148
tidal energy, 89
tipping points, 7–8
tobacco industry, 160
tornados, x
Toyota Motor Corporation, 104, 109
transmission, of electricity, 92–96
trucking, 67–68, 107, 108–10, 118, 119, 162
Trump, Donald, 45, 52, 62, 67, 151, 154; Clean Power Plan eviscerated by, 142; coal championed by, 99; energy efficiency slighted by, 66–67; Green New Deal assailed by, 156; Keystone pipeline backed by, 55, 159; methane rules rolled back by, 70; Paris Agreement abrogated by, 12, 18, 30, 42, 46, 48, 161; renewables study blocked by, 94–95
Tutos, Vanessa, 94

Uber, 148
Underdal, Arild, 48
United Kingdom, 43
United Nations Environment Programme, 9, 46, 53–54
United Nations Framework Convention on Climate Change (UNFCCC; Rio Treaty), 18–21, 30, 32, 48, 146
United Nations Sustainable Development Solutions Network, 63–64
United States: apportioning responsibility to, 127; carbon dioxide emissions in, 9, 16, 57, 103; clean technology in, 100–101, 124–25; coal mining in, 54–55, 80; emissions in, x, xiv, 1–2, 3, 9, 18, 23, 34; energy costs in, 68–69; energy sources in, 10, 15–16; fuel economy standards in, 67; international mandates resisted by, 49; net-zero status for, 34–35; Paris Agreement signed and renounced by, 1, 2–3, 18, 30, 34–35, 41, 42, 46, 48; policy blunders by, 41–42; power and leverage of, x–xi, xiv, 3, 9, 34, 46, 54, 56, 125, 140, 158; reciprocity stressed by, 41, 42, 46, 56; renewables well suited to, 87–88, 101, 124; vulnerable regions of, 7
U.S. Chamber of Commerce, 147
urease inhibitors, 71

vehicles, 33, 85; biofuel-powered, 122; buses, 107, 108, 126; efficiency gains in, 65, 140; electric, 35, 58, 62, 64, 69, 72, 91, 99, 103–11, 118, 124, 126, 141, 142, 148; emissions from, 1, 57, 67, 150; hybrid, 104; pickup trucks, 67, 103; regulation of, 1, 67; sport utility, 67, 103; trucks, 67–68, 107, 108–10, 118, 119, 162
venture capital, 61
Vibrant Clean Energy, 91

Victor, David, 34, 43, 47, 50, 55, 147
Vienna Convention (1985), 19
volcanos, 6, 137
Volkswagen Golf, 106
Volvo cars, 109

wastewater treatment, 71, 122, 123, 136
water heaters, 111, 112
water pollution, ix, 126, 153
Waxman, Henry, 144
We Are Still In, 48, 161
weatherization, 62
weather vs. climate, 4
Webber, Michael, 100, 119, 121–22
Westinghouse, 76
Whitman, Christine Todd, 24, 25
Williams, Jim, 90, 91, 131
wind energy, 35, 50, 79, 101, 155; declining cost of, 82, 84, 85, 88–91, 141; growth of, 73, 91,

96; land-based vs. offshore, 82, 83; opposition to, 160; transmission of, 94; variability of, 72, 73, 89, 92, 96–97, 131
wood, ix, 10, 116, 122
World Bank, 52
World Climate Conference (1979), 13
World Conference on the Changing Atmosphere, 16
World Meteorological Organization, 17
World Trade Organization (WTO), 50

X-energy, 77
Xi Jinping, 29, 46–47

Yucca Mountain, Nev., 78

Zero Emission Transportation Association, 148